THE WORLD OF 2044:
TECHNOLOGICAL DEVELOPMENT AND THE FUTURE OF SOCIETY

THE WORLD OF 2044:
TECHNOLOGICAL DEVELOPMENT AND THE FUTURE OF SOCIETY

Edited by

Charles Sheffield
Marcelo Alonso
Morton A. Kaplan

Paragon House

A PWPA Book
St. Paul, Minnesota

Published in the United States of America by
Professors World Peace Academy
2700 University Avenue West
St. Paul, Minnesota 55114

Trade distribution by
Paragon House
370 Lexington Avenue
New York, NY 10017

A Professors World Peace Academy Book

The Professors World Peace Academy (PWPA) is an international association of professors and scholars from diverse backgrounds, devoted to issues concerning world peace. PWPA sustains a program of conferences and publications on topics in peace studies, area and cultural studies, national and international development, education, economics and international relations.

Library of Congress Catalog-in-Publication Data

The World of 2044 : technological development and the future of society / edited by Charles Sheffield, Marcelo Alonso, Morton A. Kaplan.
p. cm.
ISBN 0-943852-49-8 : $29.95. — ISBN 0-943852-57-9 (pbk.) : $19.95
1. Technological innovations—Social aspects. 2. Technological forecasting. 3. Twenty-first century—Forecasts. I. Sheffield, Charles. II. Alonso, Marcelo, 1921- . III. Kaplan, Morton A.
T173.8.W67 1994
303.48'3'0905—dc20 94-3101
CIP

TABLE OF CONTENTS

PREFACE

Morton A. Kaplan

Few individuals have paid attention to the revolutionary changes that technology will induce not merely into society but into humans' understanding of their own humanity in the relatively near future. MTV and the sound bites of narrative TV already have changed enormously the willingness of individuals to listen to presidential debates like those of the second half of the century, let alone the kind of sustained arguments that characterized the Lincoln-Douglas debates or *The Federalist Papers*.

There is abundant evidence that technology influences our concepts of our own humanity, although sometimes it influences only the forms of the beliefs. During the period in which it was believed that witches existed, many believed that demonic creatures removed sperm from men and impregnated women. Today many believe that little gray creatures who fly around in extraterrestrial spaceships do the same thing. The advent of radio produced personality disturbances in which some individuals began to believe they were controlled by radio waves, and this indeed may become possible in the future.

Behaviorism has influenced beliefs concerning our responsibility or lack thereof for our actions. These influences may be gross or subtle. Although they may not be inescapable in the form they take, they necessarily will influence our beliefs and actions and our societies and values, including whether we permit some individuals to control others, whether through radio waves or drugs.

We believe it is ethical to grow baboons to have a supply of body parts for humans. But suppose a grossly superior extraterrestrial life form did the same thing to us. Perhaps we should consider the possibility that these ethical questions are more complicated than many scientists suppose. Scenarios of the type we discuss in this book may open our minds to ethical dimensions that will be lost if we simply allow the future to unfold.

Far more revolutionary changes are on the horizon—so revolutionary that I am willing to assert that if an individual could be placed in cryogenic sleep and awakened fifty years hence, he would find a stranger world than the denizen of the eighteenth century would have found if awakened in the early twentieth century. If this is so, obviously that is a topic of paramount importance.

Before I discuss the problem of technological projection, the projections we made, and scenarios, I should like to make my previous comment more concrete. If computers begin successfully to mimic human behavior, if prostheses begin to extend the range of human capabilities, if the genetic characteristics of fetuses can be modified, how will these developments change our understanding of our nature and our relations with each other?

Many Europeans believe that the American fascination with technology hides an indifference to humanistic studies. In the 1991 and 1992 meetings at which these papers were presented, some of the European participants criticized the topic for precisely this reason. They have forgotten their own mythology. The myth of Prometheus and even of Adam and Eve and the apple of knowledge reflect the belief that technology and information have profound consequences for human behavior and arts. This has been true throughout history but never so true as it will be in the coming generations.

Yet even the simplest aspect of these changes—that just as the 40 percent farming population of the early twentieth century has been reduced to less than percent as a consequence of new technology and information, manufacturing is headed for a similar if not identical decline as a consequence of robotics and computerization—did not percolate into the political debates of 1992. Indeed, the parties argued as if this decline could be halted or reversed. If this projection is correct, the lack of productivity improvements that currently trouble us are likely to be mere blips in the potential that will be available to us. And the downsizing for which we are currently attempting to compensate is likely to recur periodically and perhaps to turn our large corporations into obsolescent dinosaurs, again with enormous consequences for society, politics, and human values.

The current downsizing in industry—actually the United States and Japan have been less affected by job losses in manufacturing than other major nations—is a product of technology and the personnel bloating of which Parkinson wrote so eloquently in a book that was received as humor. Technological advances will replace some human functions periodically and make some skills obsolete. Personnel bloating will require contractions when competition becomes too stiff.

Already small corporations, many of which are headed by women, are producing most of the new jobs and many of the important technical innovations. As some of the technical innovations of which we shall speak come to permit an intense growth in custom designing of products in which the consumer can take the lead and in which production can be implemented almost instantly anywhere in the world, a revolution in manufacturing and the globalization of demand and supply will have occurred.

For instance, suppose the citizen of 2044 wants a new house. He or she will utilize a personal information center and scan different house designs and types. One will be selected initially. The design will be put in virtual reality mode. Furniture and living equipment will be selected and put in the house in virtual reality mode. The citizen will then walk through the house and use the furniture. Then the information center will be instructed to change the dimensions and configurations of some of the rooms. The house will be walked through and utilized again. Materials and colors will be chosen as well as alternative smart systems. All these will be priced as well, with selections available globally. Depending on design, the house will be constructed locally or robotically elsewhere and transported to site, all in a matter of days. It will be secure against attack or burglary by a variety of methods. This just skims the surface of house selection and building.

The references to education in the presidential debates did not even hint at the revolutionary changes in education that will be required if the population of the coming generations is to be usefully employed, even assuming material prosperity. The working population will require a far higher level of skill than is true today and the ability to adapt. Furthermore, creative tasks will extend far down the working ladder as the industrial privates contribute to design and business decisions. What will the effects be on self-perception and politics if the majority of the population is on the dole because it can only consume and not produce? Or even use the techniques described above without assistance? Even if the work week is drastically reduced, what will people do to make themselves feel useful? How many can be artists or creators? What dysfunctions will result? What strange religious and drug-induced stupors?

Although we did not discuss them in specific, therapeutic and biotechnological interventions will permit the enhancement of intelligence and the modification of behavior. But they will also permit forms of social control that today we would regard as evil. That was the insight that drove the dystopian scenario that I wrote.

Marcelo Alonso discusses specific technologies in his introduction to the technology chapters. I will mention just a few. New materials will permit new types of structures and designs. Alternative energy sources will likely cheapen energy and make it more efficient and less damaging to the environment, particularly in combination with new materials. Artificial intelligence, even if it does not replicate human capabilities, will permit problem solving in ways that are enormously speedier and more complex than is now possible. Transportation and communication will be enormously more efficient while virtual reality will eliminate the need for most human movement. Biotechnology and medical advances will lengthen and improve life while creating an enormous potential for misuse. New human habitats will be created and this runs the enormous risk that we may start adapting humans to environment with a potential for destroying universal human empathy.

How can we discuss such a topic even speculatively? We cannot predict the future, can we? Two years before *The World & I* published the technological projections and the initial set of scenarios, I had great difficulty convincing my editors that some types of predictions are not merely possible but highly probable.

Of course, there are clear exceptions. Even such a great physicist as James Clerk Maxwell (1831-1879) could not have predicted that enormous energy could be liberated from the atom. (It is the case, however, that Einstein, who was not familiar with the Michelson/Morley experiment, deduced the constant speed of light from Maxwellian relativity.) Perhaps other equally great surprises lie in store for us. We cannot anticipate or prepare for these, even though we may suspect that some at least will occur.

But other types of technological projections are not merely possible, but highly probable. A significant number of them are presented in this book. Of course, the year 2044 must be taken with a grain of salt. 2044 plus or minus 20 or 25 years would be more like it. Our authors may be—indeed are likely to be—wrong in some of the details. On average, however, it is highly likely that most of the technological projections will be roughly on target. And this is a matter of significant intellectual and practical interest.

Some of the counterinstances that have been cited as evidence that these projections cannot be made successfully are unconvincing. For instance, The Club of Rome surely was wrong in its projections. However, even apart from the fact that its subject matter was radically different from the technological projections made in this volume, no sooner had Meadows published the computer model that underlay the Club of Rome projections, than it was torn apart in detail by analysts at the Hudson Institute and elsewhere. That the media and many intellectuals were fooled by these projections and resistant to their disproof demonstrates only that bad science gains more public attention than good science.

The world has recently been informed by astrophysicists that sometime within the next two hundred years a particular asteroid may hit the earth with a probability of 1 in 10,000 (since corrected). This is a very small probability, but the earth-shaking catastrophe that would result warrants thought about remedial action, even today, although large-scale implementation obviously is premature. The technological changes that will be discussed in this book for the most part will happen, and during the lifetimes of more than half of those who are alive today. And these technological innovations could produce catastrophe, as well as almost utopian conditions if we adequately prepare for them.

How can we think through the social consequences of the latter possibilities? Few would have predicted the consequences of the automobile for dating, let alone for changes in the movements of humans from locus to locus and in their detachment from neighborhood. Although many predictions will be wrong and some right, that is more likely than not to be accidental, and we have no good way of choosing from among them.

This is where mind-stretching enterprises become our tools. If we cannot predict, or even project, what many of the consequences of these technological inventions may be on society and human values—although other predictions such as the decline in manufacturing jobs are predictable—we can let our imaginations roam over the possible favorable and unfavorable consequences. That at least may suggest possible consequences to be alert to and to facilitate or mitigate if they begin to appear.

The writers on technology were given a page-long list of social, political, and moral factors that should guide their choice of technologies. When the technological papers were completed, they were circulated to the scenario writers, who were to take them as the base for their imaginative scenarios. We made sure that the scenarios would range from the most optimistic to the most pessimistic. Because I wanted the reader to consider some possibilities not otherwise treated, I wrote a utopian scenario and republished an extremely pessimistic scenario that I had first written in 1965 and then, upon request from Herman Kahn, permitted him to publish as an appendix to the book he did with Anthony Weiner, *Towards the Year 2000.*

The various chapters of PWPA received the papers in advance, and several of them prepared scenarios that took into account their own cultural features. These, along with the initial set of papers, were presented at the Fifth Congress of the Professors World Peace Academy in Seoul, Korea in August 1992. The best of these are presented in the last section of this book. They at least are representative of how intelligent, educated readers from a variety of nations and cultures responded to this topic, and are intensely interesting as much for what they do not discuss as for what they do. Articles by Gregory Benford and Steven Post that were written for other audiences were added because they discussed important interfaces between biotechnology and human societies. A book review by Ben Bova from the January 1992 issue of *The World & I* was also added to present a fuller view of virtual reality as well as a scenario concerning its potential impact. Marcelo Alonso's introduction to the papers on technology gives an account of what virtual reality is.

The scenarios are so diverse in nature that there would be no point in summarizing them in this preface.[1] However, I was impressed by the fact that so many of the PWPA presentations responded to the original scenarios by stressing the importance of cultivating the family, moral values, and an ecumenical attitude toward other cultures and religions if a decent future is to be attained.

The regional scenarios generally build upon the overall projections described in the first three parts of the book. However, many of them only develop scenarios for the world of 2044 in their own regions, with only a few offering a view of the future of the world from the perspective of their region.

I was also impressed by the unwillingness of so many participants to anticipate the impact of technology upon culture, society, and politics, or even in some cases to take the coming technological revolution seriously enough to consider it. Even the best scenarios in this book, however, merely have scratched the surface. I say this not as apology or criticism, but to warn readers that this supernally important topic is likely not to be taken seriously until we invoke dangerous problems that it then may be too late to solve. Not those of production, which are relatively easy to solve, or even distribution which, if not easy, at least is dependent primarily upon political and cultural reform. But changes in how we view ourselves and our kind that may degrade what we now are, let alone what we might become. It is harder than we may believe to wake us from our contemplation of the world's navel, the present in which we are immersed.

After I had my editors on line, Glenn Strait was put in charge of acquisition. Charles Sheffield, a mathematical physicist and science fiction writer, was brought on board for the initial development of the plan for the technological papers and the scenarios. After the initial plans were formulated, and because Sheffield was too busy to handle both technology and scenarios, Marcelo Alonso, a quantum physicist, was brought in to oversee the development of the technological papers. Gordon Anderson, International Secretary General of the Professors World Peace Academy, solicited the foreign scenarios, which were intended to focus viewpoints from other than Western cultures upon scenario development. Gordon Anderson also saw the book through production.

Note

1. After reading what Jan Knappert said about predictions on the course of World War II, I could not resist this personal note. Although an undergraduate, I was scheduled to give a talk on December 9, 1941 on why the United States should enter the war. My topic was preempted by the Japanese attack on Pearl Harbor, but a large audience showed up anyway. So I decided to talk about how the war should be fought. Hitler was the most dangerous enemy, I said, so the United States should concentrate on Germany. In the meantime,' the Japanese would conquer the Philippines and Southeast Asia. A member of the audience immediately objected that we would defeat the Japanese in six weeks. The entire audience applauded and walked out on me. Draw your own conclusions about predictions from this.

PART ONE

TECHNOLOGICAL FORECASTS

INTRODUCTION

Marcelo Alonso

"The business of S&T is to discover the secrets of the Universe, and through scientific and engineering skills contribute to the security, prosperity, health and happiness of all."

— *S. Ramo, TRW*

Predicting the future has always fascinated and interested humankind. This interest has not been limited to just knowing the future but rather with the expectation of being able to influence the future in one way or another. This almost universal concern can be summarized in a single question: Can humanity choose its future? This question in turn can be split into three questions: 1) Is it possible to predict a most probable human future? 2) How does humankind, deliberate or not, shape its future? 3) Is it possible for humankind to avoid an undesirable future?

Predicting or influencing the future or evolution of a physical system is, in principle, possible. In the simple deterministic world of the French mathematician Pierre Simon de Laplace (1749-1827), the future of a physical system can be predicted with certainty if we know precisely the initial conditions and the laws regulating the evolution of the system. This is exemplified by the motions of the planets and other bodies in the solar system. However, many physical systems found on earth and all over the universe are much more complex than those known to Laplace. Their future is either unpredictable or, at best, can only be estimated probabilistically. This is the case, for example, with long-term weather forecasting, because of the extreme complexity of atmospheric dynamics.

Predicting the future of mankind in general, or of a social group in particular, is a much more difficult, almost impossible, task. Social evolution is critically affected by ideological, political, religious, and military leaders, whose appearance and subsequent actions are impossible to predict. The "human factors" affecting the development of society are so unpredictable that a comprehensive forecast of the human future must be intrinsically flawed. At best, we can guess some possible scenarios

AN UNPREDICTABLE, IRREVERSIBLE FORCE

Beyond ideology, politics, and religion, one other human factor exerts an enormous influence on social evolution in a profound, practically irreversible, and often unpredictable way. This is the human ability to acquire "knowledge" about the physical world, and to apply this knowledge to create new things that profoundly affect human life. When this unique feature of human beings is carried out in an organized and methodical way it is called "science and technology." Thus understanding the process of science and technology development provides a clue to one of the most important ways by which humankind, either deliberately or not, affects its future, as well as its present.

Looking way back in time, the list of science and technology discoveries and inventions that have produced profound social changes includes fire and agriculture, metallurgy and gunpowder, paper and the printing press, steam power and electricity. Yet none of these were the result of any desire to achieve social change. These technologies and their social effects could not have been anticipated. It would have been impossible to predict the appearance of the individuals who made the discoveries, or how the nature of their discoveries, and the way in which their discoveries were to be applied. However, many of these technologies produced social changes so profound and irreversible that we can recognize their results as "singularities in time." Even more, these early unforeseeable discoveries made civilization possible, or perhaps even inevitable. This is why the late twentieth-century Austrian economist and philosopher, Friedrich von Hayek, has said that man has been civilized against his will.

The massive mobilization of science and technology during World War II was a watershed event that later led to the development of the "postindustrial" society. This new type of society is dedicated to the systematic generation of new knowledge and its application in the economic, social, and cultural areas.

Science and technology have influenced human development in the last five decades at an unprecedented pace and scale; they are likely to exert an even more profound impact on society in the coming five decades. A few examples of social changes arising as a result of modern science and technology illustrate this fact. Radio and television have led to global communications and the reporting of events in close to real time. The creation of new hybrid grains, derived from plant genetic experiments, has permitted the "green revolution" of greatly increased crop yields and consequent major changes in the food balance between nations. The development of internal combustion engines has revolutionized transport and increased human mobility. Similar impacts could be noted for computers, nuclear energy, and pesticides.

Some science and technology achievements have also imposed serious pressure on the earth's natural resources and environment. At the same time that the "quality" of life has improved for a large portion of the world population, we have witnessed a severe deterioration of the environment and of earth's "natural" capital such as oil, minerals, soil, water, and atmosphere. The intensive use of certain technologies, the new styles of living, and the increase in population made possible by technological developments have all contributed to this environmental deterioration. Eventually, the pressure on earth resources may have serious social consequences unless some corrective measures are adopted before it is too late.

It is sobering to realize that less than 0.2 percent of the human population is engaged in science and technology, yet that small percentage has exerted and will continue to exert a major impact on human society. What should be the priority areas for scientific research? How should resources be allocated to the science and technology enterprise? Answering these questions requires that scientists and engineers work closely with political, social, and religious leaders who must provide overall guidance on what kind of future world is desirable.

Science and technology have not at all reached the limit of their possibilities in affecting society. There are many areas of research that will keep scientists and engineers busy for the next 50 years and beyond and that will have a bearing on the future of society. Thus the exercise of assessing probable technologies in advance to make sure that they help rather than harm the future of humankind is critically important.

While it would be impossible in a reasonable time to examine all probable technologies and possible human destinies that may emerge in the next 50 years, the importance of the exercise compels us to attempt it. This has been the motivation for producing this book containing articles divided between "Technological Projections" and "Possible Destinies."

"Technological" are covered in seven chapters representing six combined areas of science and technology. Our distinguished authors have committed to print their best guesstimates of how these areas will develop in the next 50 years.

Going further, and based on the technological projections, nineteen additional articles present scenarios ranging from the dystopian to the utopian.

TECHNOLOGICAL FORECASTS

A short overview of the probable technologies provides a glimpse of a vast array of developments that will surely alter the face of society. Most of the new technologies that will emerge from these research efforts will have in common that they are the result of our ever-increasing understanding of the structure of matter and our ability to manipulate it at the level of molecules, atoms, electrons and photons. As a consequence, most of the technological progress will be based on advanced computers and integrated electronic systems that operate at the atomic and electronic level.

1. **Materials.** The trend will be toward materials designed and developed based on a better understanding of atomic structure. The goals are to produce materials that are 1) lighter, 2) resistant to corrosion, 3) capable of withstanding extreme temperatures, 4) made from abundant resources, 5) easy to extract and process, 6) recyclable, 7) environmentally safe, and 8) more specialized in their use. The main materials that surely will experience further development are metals and alloys, plastics, composites, ceramics and materials with very special properties (smart materials, biomimetic, high temperature superconductivity, nanophase, nano-electronics, increased strength-to-weight ratios, etc.).

2. **Energy.** Main uses will remain generation of electric power and transportation. The goals are to develop new energy forms that are 1) reliable, 2) environmentally benign, 3) based on abundant resources, 4) safe and easy to generate and transport, 5) more efficient to use and 6)more economic. Many alternative energy sources are under intensive research but so far none of the candidates satisfies all the preceding requirements or effectively replaces the current sources (coal, gas, nuclear, hydro, and solar). The main developments will be in 1) photovoltaics and non imaging solar concentrators, 2) advanced nuclear reactors (fission reactors with inherent safety features and modular design, improved breeder and

converter reactors, fusion reactors), 3) high temperature superconductors, 4) energy storage, and 5) space power satellites beaming to Earth solar energy as microwaves, 6) fuel cells, 7) reliable energy storage systems, and 8) transmission lines using superconductors.

3. **Robotics and Artificial Intelligence.** The goal is to develop "expert systems" capable of receiving, storing and processing data and learning as well as evaluating situations and reacting accordingly (recognize, reason, talk, decide, act). The basic requirement is to mimic nervous systems and behavioral features of humans and animals, for which faster and more powerful computers (above 100 GHz) are needed. An extension of artificial intelligence called "intelligence amplification" is discussed below.

4. **Transportation.** The goals are to 1) increase speed, 2) reduce congestion, 3) increase safety, 4) increase efficiency, 5) reduce ratio of transportation cost to that of goods and services, 6) reduce pollution, and 7) reduce need to move people. No major developments are expected but the main trends are 1) combination of electric and internal combustion engines in cars, 2) smart cars with integrated electronic systems for monitoring engine and cruise control and with navigational sensors, 3) smart highways that automatically direct and regulate the flow of vehicles, 4) air traffic and collision avoidance systems using transpondents that broadcast enhanced radar echo, 5) new airplanes (STOL, VTOL, SST) that may be unmanned for freight transport, 6) maglev trains circulating in vacuum tubes, 7) combined rapid transit systems for area-to-point-to-point-to-area, and 8) new vehicle designs based on innovative materials and processes. New developments in the making are technologies for space transportation. Candidates are single-stage-to-orbit (SSTO) systems (Delta-Clipper or DC-X and Space Ship Experimental or SSX vehicles) and the National Aero Space Plane (NASP), with advanced rocket and ramjet-scramjet engines.

5. **Biotechnology.** Developments will be based on improved understanding of living systems at the molecular and genetic levels, particularly in relation to the structure, functions and manipulation of genetic material and bio-molecules (enzymes, proteins, antisense RNA, etc.). Main areas of development are 1) technologies for gene transfer, repair, inhibition, enhancement, and splicing, and 2) control of enzymatic processes (catalytic antibodies, hybridomas). Some applications will be 1) regulation of population growth by hormonal control of reproductive functions, 2) increase of food production by developing transgenic varieties (plants, fish) and cell cultures, intensive agriculture (less water, energy, fertilizers, pesticides), 3) improve life span in healthier conditions, and 4) increase biomass fuels (fast growing trees and shrubs) reducing CO_2 in the atmosphere and increasing CO intake by green plants (algae in oceans and lakes).

6. **Medical Technology.** The goal is to reduce dependence on the "physician factor" (personal knowledge, experience, information, observation, therapy) by the "instrument factor" using advanced physical techniques, and increase the use of computers for more resolution, sensibility, reliability, precision and speed, both in diagnostics and therapy. Developments in diagnostics include 1) ultrasounds, 2) lasers and fiber optics, 3) non invasive probes, 4) imaging, and 5) expert systems. Therapeutic developments will include 1) drug delivery systems controlled bio-electronically, by microrobots, or by the central nervous system, 2) minimal invasive surgical techniques (lasers, microrobots), 3) robotic prosthesis controlled by nervous system, 4) organ and tissue transplantation (xenografts, biohybrids), 5) reproductive techniques (cryogenic preservation of embryos, embryo splitting) 6) remote control of human motion through access to the central nervous system (access to nerve impulses, decoding of nervous signals), 7) organ replacement using bionic materials, and 8) diagnosis of diseases and their

treatment by genes analysis and replacement. An even more dramatic advance in the next decades will be that of "neural prosthesis," interfacing the nervous system with machines; employing a television camera attached to tiny electrodes feeding into specific areas of the brain, it will be possible for the blind to see, the deaf to hear, and the crippled to walk. Conversely, through recording electrodes attached to the brain and a special radio transmitter, it will be possible to send signals from the brain's motor cortex to operate computers, typewriters, and other devices, just by using the brain.

7. **New Human Habitats.** The objective is to move people from coastal areas, either inland, out to the sea or out into space. *Ocean habitats* would be floating cities built downward where the temperature remains practically constant (eco-baths). Main activities would be ocean bottom mining, fishing support, ocean farming, marine protein processing, desalination, waste conversion, oceanographic research, assembly, packing and processing industries, radar and weather stations, and resort centers. *Space habitats* would consist of rotating spherical shells of metal and glass, several kilometers in diameter, enclosing a normal breathable atmosphere, using clean solar energy and moon and asteroids materials, capable of supporting 10,000 inhabitants. Gravity is produced by rotation and the sphere would be placed about two-thirds of the distance from earth to the moon. Space habitats would serve to produce microwaves (2.4 Ghz) from solar energy (90 percent efficiency) in Solar Power Satellites (SPS) and beam them to collecting antennas on earth. They would be subsequently transformed into electric power that could be sent to different places on Earth. Space habitats could support some industries and closed-cycle agriculture.

8. **Communications.** The world will become more interconnected through a global communication system for fast data processing and transmittal and for enhanced land, sea and air navigation and communication (electronic

highways). Interactive systems will combine video and smart computers as well as advanced software and animation. Integrated information systems will combine telephone, video and fax, reducing the need for people to travel. All these developments will depend on more advanced microelectronics and chips, optical fibers and an extensive network of satellites.

9. Intelligence Amplification. Each of these areas is analyzed in detail in the following chapters. However, there is an area still under development that has not been considered explicitly in those chapters but that surely is going to have a great impact on society in the decades to come. It is that of advanced computer technologies. Since their first phase of development, computers have been used primarily as number crunchers handling mathematical operations that are either too complex, too elaborate, or too difficult to be done by hand, or for performing, in a dramatically efficient way, tasks such as word processing and graphics that can be programmed using the algorithm of Boolean algebra. But with recent developments in speed and capacity of microprocessors and memory chips a new role of computers as "mind amplifiers" is emerging that will allow extending human intellectual capabilities. This new role is called *intelligence amplification* (IA), that is different from artificial intelligence (AI). In AI the idea is to design machines that, with proper programming and data base, can perform the same tasks as the human brain (or mind) and body to the extent that they can replace humans in many instances; this is the case of robots. The objective of IA has more profound implications: the design of systems based on computers and peripherals, that amplify the human mind by doing things that the human mind cannot do or has trouble doing.

The human mind excels any existing computer system in three areas. One is pattern recognition, another is evaluation of sensorial experiences as they are received, and the third is context recognition, by which past experiences are correlated with present ones. On the other hand, modern computer systems excel the human mind in computations, in handling and correlating large amounts of information including graphics, and in storing information and remembering it precisely when needed.

Therefore, by improving the human-computer interface it is becoming more and more feasible to carry out tasks too difficult or even impossible for the human mind. For example, it is already possible to use computers for analyzing the structure of complex molecules and their forms of binding, for analyzing industrial and architectural designs from different angles before actual construction, for the visualization in real time of the functioning of the human body, that is "medical imaging," and for following the evolution of a hurricane.

However, current computer systems have several limitations in providing a faithful representation of reality. These systems provide only a two-dimensional (2D) visual representation with, at its best, a limited audio component. However, our perception of reality is three-dimensional (3D), both visually and acoustically, and involves practically all our senses, giving a feeling of textures, distance, temperature, odor, etc., as well as of our position relative to the environment. Therefore, the challenge faced by designers of computer systems such as IA is to duplicate electrically, optically, acoustically and mechanically the sensory information that constitutes our perception of reality. This imposes severe technological requirements, such as faster processing and much larger RAM, and needs a much more elaborate human-computer interface than what is available now. The product of the new IA systems is what has been called *virtual reality* (VR) or *cyberspace* (from the Greek word "cyber," meaning "steersman"). VR systems must consist of a 3D computer-generated representation of a real space with total sensorial immersion of the user through the proper combination of different sensory transducers. The user then perceives a real space without leaving his place in front of the computer.

Several partial solutions to the design of VR systems already exist. Flight simulators were one of the first VR systems in which the pilot can practice many flying tasks while sitting safely on the ground, but simulators have only limited visual and acoustical sensing. Another limited VR system has been cinerama, that provides a 2D image but with a larger visual field than ordinary screens and that approximates the normal visual field of a person, coupled with surround sound. Surround sound is already in use for home sound and video reproduction

One of the most difficult tasks for VR is to produce a truly 3D representation that not only appears 3D but that changes as the observer changes his line of vision, as happens with real objects, coupled with full sensorial immersion. One of the first attempts to produce a VR system has been head-mounted devices (HMD), initially developed at Massachusetts Institute of Technology and Harvard and later on at the University of Utah, with the sponsorship of ARPA and ONR. HMD consists of a helmet with special goggles and earphones for 3D visual and sound reception, as well as electronically activated gloves and other sensors, creating the feeling of presence, of being there. But HMD are complex (composed of several coupled subsystems capable of intensive mathematical translation of the user's position and room coordinates), cumbersome to use, and expensive so that their widespread use in their present form is very limited. HMD systems are under intense research and if successfully developed in the next two or three decades they may have several applications in the ways people communicate and exchange information (telecommunication), including conducting meetings in cyberspace. Another recent development is the "cyberspace communicator," being developed in Japan, that can produce a 3D visual representation without the need of an HMD. Similarly "exoskeletal controllers" under development might produce a whole-body sensorial interface with VR devices.

There is no question that these new VR telecommunication technologies will have an impact on how people see reality (think, perceive, believe). VR will also change how scientists carry out their research, how industrial engineers and architects design their processes, plants and buildings, how medical practitioners deal with their patients, how artistic expressions (theater and music) are received at home. VR systems, coupled with robots, might be used for several industrial and medical purposes. It is even possible that quadriplegics might gain a freedom that their bodies do not provide. And so on.

There is no question that a "new brave world" based on VR is in the making. The big question is if this emerging symbiotic relation between humans and computers can contribute to make this new world a better one. The hope is that the answer is in the positive.

CONCLUSION

The magnitude of the global S&T effort to be carried out in the next decades shows that the S&T complex should be considered as an industry that consumes resources (human, financial, material) and produces diverse results (ideas, know-how, materials, processes, etc.). In this respect, the S&T complex responds to market pressures and transcends the domain of the individual practitioner, becoming a matter of collective social interest and concern. That implies the need to manage S&T efforts at the multinational, national, corporate and institutional levels to make sure that societal needs are properly satisfied worldwide and without curtailing the initiative of the individual scientist. Only through proper management of the S&T complex, without hindering free scientific inquiry, can society hope that S&T will be able to help choose an acceptable future.

However, it should be reiterated that the course of the world's development will not be determined by S&T alone but rather through the interplay of science and technology with social and political programs, with religious influences, with natural as well as manmade events, and with other concurrent situations. In addition although most S&T innovations are developed by the more technologically advanced countries, they, in general, have a global impact, affecting all countries regardless of their level of development. That makes even more critical exercises like this one to try to assess in advance the trends in S&T efforts in order to make sure that they help rather than harm the future of mankind. This also poses a serious responsibility on scientists and technologists, who constitute less than 0.2 percent of the world population but whose actions may dramatically and massively affect, in one way or another, the whole world.

ONE

MATERIALS AND ENERGY

Alexander Zucker

Materials define the limits of technology. This is particularly true in the case of the diverse energy technologies, where properties of materials affect the efficiency and reliability of machinery and processes that convert energy to perform work, or make products that suit our purposes. In every sector of energy—including transportation, heating and cooling, industry, and agriculture—new materials are being developed. This may not be surprising in newly discovered fields such as super-conductivity, but it is happening with equal vigor in the age-old art of steelmaking. And no wonder; the demands placed on energy technologies are truly extraordinary. Not only will energy demand grow everywhere in the world, most of all in the developing nations, but also, as environmental issues increasingly dominate the human use of technology, greater emphasis will be placed on environmentally benign and much more efficient energy machinery. This, in turn, means that materials in the future will have to be lighter, more resistant to corrosion, better able to withstand the rigors of high temperature service, and made of substances plentiful in nature and relatively easy to extract.

Steel will continue to dominate the energy materials field. It is by far the cheapest, most plentiful, and most versatile material we have. But it will not be the same steel we are used to now, nor will it be made in the same way. Microalloying (the adding of tiny amounts of special elements) is the steelmakers' game of the future. It is now known that mixing very low levels of additives with steel can make a big difference in its properties. Adding one-tenth of a percent (0.1%) of niobium in a steel alloy can have major effects on its strength, on its ductility (the ability to deform without breaking), and its behavior at high temperatures.

Steelmaking, in fact and fable, is a symbol of the passing age. It has the romance of giant technology: huge cranes lifting 50-ton ladles of molten metal, 400 tons of glowing liquid metal pouring out of a tipping furnace, sparks flying, mile-long rolling mills on which the sheet steel is worked as it moves at speeds of 50 feet per second. An industry vertically integrated from its iron ore mines, coal mines, barges, coke ovens, and huge steel mills, it employs thousands of people. It is no longer dirty nor is its hallmark any longer the smell of acrid smoke, but it is still very energy-intensive every step of the way.

In the next 50 years steel will probably remain at the top of the materials list, at least as far as tonnage is concerned, but it will spawn a totally different industry, Most of the steel will be recycled scrap. And it is about time! Here are some current figures for the United States: Annually we produce about 99 million tons of new steel, of which 45 million tons is recycled scrap. The prospect is that our steel industry, by halfway through the next century, will be largely devoted to recycling scrap, adding at most 10 million tons of new steel, making steel better, longer lasting, and tailored to its use through microalloying. Steel will also be made differently. The 500-ton furnaces will be gone, replaced by processes that make fine powders that are then sintered into shapes needing very little further machining. Energy consumption will decrease dramatically, and because of the intensive use of scrap, environmental problems due to iron and coal mining will be much alleviated.

Steel will remain with us, but it will have to fight to maintain its market. Hard on its heels will be the lighter and more plentiful metals, aluminum and magnesium. Consider, for example, an alloy containing 25 atomic percent aluminum and the rest largely iron and small amounts of elements like boron and chromium. The alloy turns out to be much stronger than many of the so-called super alloys and superior to stainless steel in its resistance to corrosion. Magnesium too is an untapped metallurgical treasure. It is much lighter than aluminum and can be made strong and durable by alloying. However, it has two drawbacks: It burns readily with a very hot flame, and it corrodes easily. Fifty years seems long enough for metallurgists to find ways around these weak points. Assuming appropriate extractive technologies, there is a nearly inexhaustible supply of magnesium in the oceans, so we can confidently look forward to lighter and more fuel-efficient cars, trucks, airplanes, and ships.

Steel will also have to compete with other materials. Foremost among them are plastics and composites. Plastics are technically polymers, materials made of chains of interconnecting long, twisted molecules, while composites are made of several materials usually contained in a matrix of one of them. Common examples of plastics are polystyrene or polyurethane; composites are frequently fiber reinforced materials such as fiberglass used in boats, or carbon-reinforced materials used in golf clubs. Plastics and especially composites are getting stronger, and strength is where steel had the territory to itself for such a long time. Plastics are also extending their useful temperature range, so more of them will be used not only in car bodies but also under the hood. Again, lighter materials save energy. Energy conservation, in turn, benefits the environment and the pocketbook.

Close behind plastics are the emerging developments in ceramics. Here is the real curiosity. Ceramics are civilization's oldest materials. Yet they are also high tech's newest. Spurred largely by their strength and unbeatable high temperature properties, the United States, Japan, and Germany are investing heavily in ceramics research. By the middle of the next century we can expect an all-ceramic automobile engine. It will be small and light, it will cool itself through the exhaust, and because it will operate at much higher temperatures it will also be much more energy efficient. In 2044, when you raise the hood of your synthetic-methanol-fueled car, you will see a small hot ceramic engine, no cooling system, and no emission controls. With luck and the application of all the weight-saving materials, you may well expect 50 miles per gallon (gasoline equivalent) and good acceleration in an uncramped car.

What has happened and what needs to keep happening to guarantee that new materials will be here when we need them?

Over the last 20 years a quiet revolution has taken place in materials science. It is based on the development of new instruments that allow us to look at materials in ways no one had dreamed possible a generation ago. Some have familiar names like electron microscope or X-ray camera, but the majority of the instruments and techniques are new even to the initiated: low energy electron diffractometers, imaging atom probe, small angle neutron scattering, positron annihilation, atomic resolution microscopes, and more.

This vast armamentarium of expensive new instruments (they cost about a million dollars apiece) has one ultimate goal: to observe materials at the atomic scale, that is, to see the position of different atoms and to identify them, to deduce from these observations the atomic structure on scales that vary from collections of a few atoms to many millions of atoms. This information is them coupled with our theoretical or empirical understanding, and with the aid of large modern computers, models are developed that provide deeper understanding as well as reliable predictive capabilities. In principle it is all simple enough: Structure defines properties; therefore, an understanding of structure will predict properties. In practice things are not nearly so simple, and while many important advances have been made, the science of materials is just entering its productive phase.

The matter becomes even more complex because prediction of properties is not enough; to be of any use, materials have to be produced in sizable quantities by means of economical and practical processes.

We are then faced with a triad of challenges. To bring out a material with certain desirable properties—be that an improved stainless steel, a better solar photoelectric cell, or a useful high-temperature superconductor—we have to study and ascertain its structure, deduce how that structure leads to the properties in question, and finally develop an industrial process that will manufacture the material with the right structure and the right properties at the right price.

It is daunting, but in a few cases it is now possible. The work of metallurgists, ceramists, and condensed matter physicists for the next half century is then pretty well laid out. Just imagine the joy and excitement of using not only those new machines to probe matter in exquisite detail, but also the beautifully complex equations of quantum mechanics that describe matter so elegantly. The phrase "mind over matter" was really coined for this wonderful prospect.

Leaving the domain of structural materials—steel, light alloys of aluminum and magnesium, ceramics, and plastics—we enter the world of electronic materials, the real miracles of the last half of the twentieth century.

"Photovoltaics" is the technical name for solar cells that convert sunlight to low-voltage electricity. They are now mostly made of silicon, although more exotic, and naturally more expensive, materials like gallium arsenide also have some desirable properties for special applications. The best silicon solar cells now operate at efficiencies of about 20 percent, that is, one-fifth of the energy that hits the surface is converted to electricity. These cells are special in that they consist of a single crystal of silicon, just as the diamond in an engagement ring is a single crystal of carbon. It would cost less to make solar cells of polycrystalline silicon, but then the efficiency drops to less than 10 percent. The cost of photovoltaics is usually expressed in dollars per installed watt. Fifteen years ago, during the energy frenzy that followed the formation of the Middle Eastern oil cartel, an installed photovoltaic watt cost about $50. With the current generation of cells, it now stands at $5. By way of comparison, the capital costs of large coal-fired or nuclear power plants are $1-2 per watt.

Photovoltaics have taken giant strides, mainly because we now understand them better theoretically and because processing technology has been much improved. But the practical use of solar photovoltaics must contend with another serious obstacle, namely, energy storage. The sun does not shine all day, or even every day. And while improvement of solar cells is clearly a matter for materials scientists, so is the problem of storage. What is needed is a vastly improved battery that is cheap, able to hold a much larger charge per unit volume than, say, a lead acid automobile battery, that can be discharged and charged thousands of times, and that is easy to operate and maintain. There is no such thing now. But there are other ways to store energy. An attractive alternative would be to electrolyze water (convert it to molecular hydrogen and oxygen) using electricity generated in photovoltaic cells that absorb solar energy while the sun is shining, and store the resulting hydrogen. Some metals like titanium can store huge amounts of hydrogen safely and at low pressure, but after a number of cycles of storage and extraction, titanium's ability to absorb hydrogen is much diminished, for reasons that are not clear at this time. So there is another materials problem!

The future of photovoltaics, however, is not without hope. In 50 years we will almost surely see better and cheaper solar cells, perhaps made of new organic semiconductors that can be mass produced and used in special circumstances. Storage problems are more intractable, although hydrogen storage in one form or another can probably be mastered. Will cheap, safe, plentiful solar energy power the world of our grandchildren? I doubt it.

The most astonishing electronic materials story is that of superconductivity. The phenomenon that some materials at temperatures very close to absolute zero lose all resistance to an electric current passing through them was discovered in 1911, nearly a century ago. However, the correct quantum mechanical explanation was not worked out until 1957. That theory, in turn, failed to predict one of the most tantalizing discoveries of the last decade: high-temperature superconductivity (HTS). This new observation, which still defies explanation by theory, seems to be only distantly related to the original, or low temperature, superconductivity. But then, what a splendid class of technological dreams can be induced by HTS. All electrical machinery, generators, and motors shrink by orders of magnitude in size, become wildly efficient, and can be used in as yet unimagined ways. Strong magnetic fields using super-conducting magnets become practical, to be used in levitated trains in urban and intercity transport, in fusion reactors to contain the million-degree plasma, or in HTS low-voltage transmission lines to transmit power from distant generating plants to the urban areas where it will be used.

The HTS material, however, is really special. It is a layered compound, surprisingly easy to make, of yttrium, copper, barium, and oxygen, or a few variations on the theme. And it is superconducting; it presents no resistance to electric current, at temperatures as high as 120 Kelvin. What is so magic about that temperature? Simply that it is above 78 Kelvin, the boiling point of liquid nitrogen, which is cheap, safe, and easy to store and to make. But there is more to consider. Present-day HTSs can carry only very limited amounts of electricity; exceed them and the material ceases to be a superconductor. Another problem: HTSs are ceramics and as such not easily wound into coils or other shapes dear to electrical engineers. But there is no reason why in the next 50 years the world's material scientists should not (a) bring about a fundamental understanding of the HTSs, which will (b) result in our ability to synthesize other and even more attractive HTSs, which in turn will (c) change entirely the way we use electricity. What a charming prospect: the kitchen fridge could dispense not only ice and ice water but also liquid nitrogen to keep all those household motors humming.

Science and technology are optimistic human enterprises. By their very nature they promise a better life in the future: better health, more leisure, faster and cheaper travel, more information available to us, and so on. When one sphere faces limits, another opens up with apparently limitless possibilities. If Malthus' predictions of certain starvation, poverty, and disease has not materialized, we have science and technology to thank for it: science for opening our eyes to many of nature's secrets, and technology for applying new knowledge to the service of humanity. Are there not merciless Malthusian limits to the use of nonrenewable resources, constrained as they are by their finiteness in the earth's crust? That is a good question, and must be dealt with by anyone purporting to look into that cloudy crystal ball that holds the future of civilization.

Before we can meaningfully ask whether materials resources will run out, we have to postulate a model of the future world society. A conservative estimate of the population by the year 2044 would be 8.5 billion people. On the other hand, we are optimistic about economic growth. World materials consumption can be based on the premise that over the next century the per capita demand for materials in the developed countries will remain constant, and that per capita demand in the developing countries will grow from its present value of 6 percent of that for developed countries to 50 percent, roughly equivalent to that of present-day Greece or Portugal.

Based on these assumptions, there are many substances or elements in virtually infinite supply, and others that will be exhausted in the course of the next century. For example, we can obtain a seemingly unlimited amount of oxygen and nitrogen from the air, sodium and magnesium from the oceans, and aluminum, silicon, calcium, and a few others from the earth's crust. Iron, which is so crucial to all aspects of energy conversion and use, is a special case. Current resources, including taconites with more than 15 percent iron, should last for centuries, and many more centuries' supply can be furnished from ultrabasic rock.

But the good news pretty well ends there. Copper will run out in the middle of the next century, lead will disappear even sooner (so a lead acid battery for electric cars is not a long-term option, but a sodium-sulfur battery is). All known HTSs contain copper, so the shortage of copper may limit their applications. Steel alloying elements like manganese, molybdenum, and nickel will be gone, but chromium will not be. Silicon, the quintessential high-tech material, is virtually infinite in the world's sands, but germanium will run out, as will zinc, tin, and bismuth. One way to stretch the availability of these nonrenewable materials is through technologies for conservation and recycling.

The world's resources will not run out all at once. When resources become scarce, their price goes up. This stimulates more prospecting and recovery from less abundant sources. There may even be a temporary glut and a drop in price. Eventually and inexorably, however, even the poorer deposits become exhausted, the price goes up, and then a whole new thing happens: A substitute is found, either for the material or for the technology based on that material. More than ever the future materials scenario calls for a lively, well-supported scientific enterprise that can come up with the right answer when a shortage threatens—that coupled with a conservative ethic—conservative not in politics but in resources. An ethic that will demand products that last, that are designed to be repaired instead of thrown away, and if they have to be discarded they will reappear from scrap to serve us anew. Human ingenuity can surely triumph again over the Malthusian monster, but only if man adopts an ethic that preserves the earth in all its aspects, nurturing it as it has for so long nurtured us.

TWO

ROBOTICS AND ARTIFICIAL INTELLIGENCE

Hans Moravec

Instincts regarding the nature and quantity of work we enjoy probably evolved during the 100,000 years our ancestors lived as hunter-gatherers. Less than 10,000 years ago the agricultural revolution made life more stable, and richer in goods and information. But, paradoxically, it requires more human labor to support an agricultural society than to live in a primitive one, and the work is of a different, "unnatural," kind, out of step with old instincts. The effort to avoid it has resulted in domestication of animals, slavery, and the industrial revolution. Many jobs must still be done by hand, engendering for hundreds of years the fantasy of an intelligent but soulless being that can tirelessly dispatch the drudgery.

But only in this century have electronic sensors and computers given machines the ability to sense their world and to think about it, and so offered a way to fulfil the wish. As in the fables, the side effects are likely to dominate the resulting story. Most significantly, these perfect slaves will continue to develop, and will not long remain soulless. As they increase in competence they will have occasion to make more and more autonomous decisions, and so will slowly develop a volition and purpose of their own. At the same time they will become indispensable. Our minds evolved to store the skills and memories of a Stone Age life, not the enormous complexity that has developed in the last ten thousand years. We've kept up, after a fashion, through a series of social inventions—social stratification and a division of labor, memory aids like poetry and schooling, written records stored outside the body, and recently machines that can do some of our thinking entirely without us. The portion of absolutely essential human activity that takes place outside of human bodies and minds has been steadily increasing. Hard working intelligent machines may complete the trend.

Serious attempts to build thinking machines began after World War II. One line of research, called cybernetics, used simple electronic circuitry to mimic small nervous systems, and produced machines that could learn to recognize simple patterns, and turtle-like robots that found their way to lighted recharging hutches. An entirely different approach, named Artificial Intelligence (AI), attempted to duplicate rational human thought in the large computers that appeared after the war, and by 1965 had demonstrated programs that proved theorems in logic and geometry, solved calculus problems, and played good games of checkers. In the early 1970s, AI research groups at the Massachusetts Institute of Technology (MIT) and Stanford University attached television cameras and robot arms to their computers, so their "thinking" programs could begin to collect their information directly from the real world.

What a shock! While the pure reasoning programs did their jobs about as well and about as fast as college freshmen, the best robot control programs took hours to find and pick up a few blocks on a table, and often failed completely—a performance much worse than a six-month-old child. This disparity between programs that reason and programs that perceive and act in a real world holds to this day. At Carnegie Mellon University there are two desk-sized computers than can play chess at grandmaster level, within the top 20 players in the world, when given their moves on a keyboard. But present-day robotics could produce only a complex and unreliable machine for finding and moving normal chess pieces themselves.

In hindsight it seems that, in an absolute sense, reasoning is much easier than perceiving and acting—a position not hard to rationalize in evolutionary terms. The survival of human beings (and their ancestors) has depended for hundreds of millions of years on seeing and moving in the physical world, and in that competition large parts of their brains have become efficiently organized for the task. But we didn't appreciate this monumental skill because it is shared by every human being and most animals—it is commonplace. On the other hand, rational thinking, as in chess, is a newly acquired skill, perhaps less than one hundred thousand years old. The parts of our brain devoted to it are not well organized, and, in an absolute sense, we're not very good at it. But until recently we had no competition to show us up.

By comparing the edge- and motion-detecting circuitry in the four layers of nerve cells in the retina, the best understood major circuit in the human nervous system, with similar processes developed for "computer vision" systems that allow robot in research and industry to see, I've estimated that it would take a billion computations per second (the power of a world leading Cray 2 supercomputer) to produce the same results at the same speed as a human retina. By extrapolation, to emulate a whole brain might take ten trillion arithmetic operations per second, or ten thousand Crays worth. This is for operations our nervous system do extremely efficiently and well.

Arithmetic provides an example at the other extreme. Last year a new computer was tested for a few months with a program that computed the number π to more than one billion decimal places. By contrast, the largest unaided manual computation of π was 707 digits by William Shanks in 1873. It took him several years, and because of a mistake every digit past the 527th was wrong! In *arithmetic* today's average computers are one million times more powerful than human beings. In very narrow areas of *rational thought* (like playing chess or proving theorems) they are about the same. And in *perception* and *control of movement* in the complex real world, and related areas of common sense knowledge and intuitive and visual problem solving, today's average computers are a million times less capable.

The deficit is evident even in pure problem solving AI programs. To this day AI programs exhibit no shred of common sense—a medical diagnosis program, for instance, may prescribe an antibiotic when presented a broken bicycle because it lacks a model of people disease, or bicycles. Yet these programs, on existing computers, would be overwhelmed were they to be bloated with the details of everyday life, since each new fact can interact with the others in an astronomical "combination explosion." (A ten-year project called *Cyc* at the Microelectronics and Computer Consortium in Austin, Texas is attempting to build just such a common sense data base. They estimate the final result will contain over one hundred million logic sentences about everyday objects and actions.)

Machines have a lot of catching up to do. On the other hand, for most of the century, machine calculation has been improving a thousandfold every twenty years, and there are basic development in research labs that can sustain this for at least several decades more. In less than fifty years computer hardware should be powerful enough to match, and exceed, even the well-developed parts of human intelligence. But what about the software that would be required to give these powerful machines the ability to perceive, intuit and think as well as humans? The Cybernetic approach that attempts to directly imitate nervous systems is very slow, partly because examining a working brain in detail is a very tedious process. New instruments may change that in the future. The AI approach has successfully imitated some aspects of rational thought, but that seems to be only about one millionth of the problem.

I feel that the fastest progress on the hardest problems will come from a third approach, the newer field of robotics, the construction of systems that must see and move in the physical world. Robotics research is imitating the *evolution* of animal minds, adding capabilities to machines a few at a time, so that the resulting sequence of machine behaviors resembles the capabilities of animals with increasingly complex nervous systems. This effort to build intelligence from the bottom up is helped by biological peeks at the "back of the book"—at the neuronal, structural, and behavioral features of animals and humans.

The best robots today are controlled by computers just powerful enough to simulate the nervous system of an insect, cost as much as houses, and so find only a few profitable niches in society (among them, spray painting and spot welding cars and assembling circuit boards). But those few applications are encouraging research that is slowly providing a base for a huge future growth. Robot evolution in the direction of full intelligence will greatly accelerate, I believe, in about a decade when the mass-produced general purpose, *universal*, robot (which I will also refer to as the "Volks-Robot," from the German "people") becomes possible.

THE VOLKS-ROBOT (ca. 2000-2010)

The first generation of universal robot will need to navigate reliably and safely over stairs and rough and flat ground. It must be able to manipulate most objects, and to find them in the world in front of it. There are beginnings of solutions today. Hitachi of Japan has a mobility system of five steerable wheels, each on its own telescoping stalk that allows it to accommodate to rises and dips in the terrain, and to raise one on a stair, while standing stably on the other four. My laboratory at Carnegie Mellon University in Pittsburgh has a navigation method that allows a robot equipped with sonar range measuring devices and television cameras to build probabilistic maps of its surroundings to determine its location and plan routes. An elegant three-fingered mechanical hand at MIT can hold and orient bolts and eggs and manipulate a string in a humanlike fashion. A system called 3DPO from Stanford Research Institute International in Menlo Park, California can find a desired part in a jumble seen by special range-finding camera. The slow operation of these systems suggests one other element needed for the Volks-robot, namely a computer about one thousand times as powerful as those found on desks and in robots today. Such machines, able to do one billion computations per second, would provide approximately the brain power of a mouse (with different abilities).

Universal robots will find their first uses in factories, where they will be cheaper and more versatile than the older generation of robots they replace. Eventually they will become cheap enough for some households, extending the reach of personal computers from a few tasks in the data world to many in the physical world.

As with computers, many applications of the Volks-robots will surprise their inventors. Some will do light mechanical assembly, clean bathrooms, assemble and cook gourmet meals from fresh ingredients, do tuneups on a certain year and make of cars, hook patterned rugs, weed a lawn, run robot races, do detail earthmoving and stonework, investigate bomb threats, deliver to and fetch from warehoused inventories, and much more. Each application will require its own original software (very complex by today's computer program standards), and some may also need optional hardware attachments for the robot such as special tools and chemical sensors. Robots will also acquire new abilities: learning, imagining, and reasoning.

LEARNING (2010-2020)

Useful though they will be, the first generation of universal robots will be rigid slaves to simple programs. If the machine bangs its elbow while chopping beef in your kitchen making stroganoff, you will have to find another place for the robot to do its work, or beg the software manufacturer for a fix. Second generation robots with more powerful computers will be able to host a more flexible kind of program able to adjust itself by a kind of conditioned learning. In other words, they will be adaptive robots. First generation programs will consist primarily of sequences of the type "Do step A, then B, then C...." The programs for the second generation will read "Do step A1 or A2 or A3... then B1 or B2 or B3... then C1 or C2 or C3...." In the beef stroganoff example, A1 might be to chop with the right hand of the robot, while A2 is to use the left hand. Each alternative in the program has a "weight," a number that indicates the desirability of using it rather than one of the alternate tasks. The machine also contains a "pain" system, a series of programs that look out for problems, such as collisions, and respond by reducing the weights of recently invoked branches, and a "pleasure system that increases the relevant weights when good conditions, such as well charged batteries or a task efficiently completed, are detected. As the robot bangs its elbow repeatedly on a microwave in your kitchen, it gradually learns to use its other hand (as well as adapting to its surroundings in a thousand other ways). A program with many alternatives at each step, whose pain and pleasure systems are arranged to produce a pleasure signal on hearing the word "good" and a pain message on hearing "bad" could be slowly trained to do new tasks, like a dog or a cat.

IMAGERY (2020-2030)

Adaptive robots will find jobs everywhere, and the hardware and software that supports them could become the largest on earth. But teaching them new tasks, whether by writing programs or through punishment and reward, will be very tedious. This deficiency will lead to a prodigious innovation, a software *world-modeler* (requiring another big increase in computer power), that allows the robot to simulate its immediate surroundings and its own actions within them, and thus to think about its tasks before acting. Before making beef stroganoff in your kitchen, the new robot would simulate the task many times. Each time its simulated elbow bangs the simulated cabinet, the software would update the learning weights just as if the collision had physically happened. After many mental run-throughs the robot would be well trained, so that when it finally cooks for real, it does it correctly.

The simulation can be used in many other ways. After a job, the robot can run through its previous actions and try variations on them to improve future performance. A robot might even be configured to invent some of its own programs by means of a simpler program that can detect how nearly a sequence of robot actions achieves a desired task. This training program would, in repeated simulations, provide the "good" and "bad" indications needed to condition a general learning program like the "dog or cat" program of the previous decade.

It will take a large community of patient researchers to build good simulators. A robot entering a new room must include vast amounts of prior knowledge in its simulation, such as the expected shapes and probable contents of kitchen cabinets and the effect of (and force needed for) turning faucet knobs. It needs instinctive motor-perceptual knowledge about the world that took millions of years of evolution to install in us, that tells us instinctively when a height is dangerous, how hard to throw a stone, or if the animal facing us is a threat. Robots that incorporate it may be as smart as monkeys.

REASONING (2030-2040)

In the decades while the "bottom-up" evolution of robots is transferring the perceptual and motor faculties of human beings into machinery, the conventional Artificial Intelligence industry will be perfecting the mechanization of reasoning. Since today's programs already match human beings in some areas, those of 40 years from now, running on computers a million times as fast as today's, should be quite superhuman. Today's reasoning programs work from small amounts of clear and correct information prepared by human beings. Data from robot sensors such as cameras is much too voluminous and too noisy for them to use. But a good robot simulator will contain nearly organized data about the robot and its world—for instance, if a knife is on the countertop, or if the robot is holding a cup. A robot with simulator can be married to a reasoning program to produce a machine with most of the abilities of a human being. The combination will create a being that in some ways resemble us, but in others are like nothing the world has seen before.

FIRST GENERATION TECHNICALITIES

Both industrial robot manipulators and the research effort to build "smart" robots are twenty-five years old. Universal robots will require at least another decade of development, but some of their elements can be guessed from the experience so far. One consideration is weight. Mobile robots built to work in human sized spaces today weigh too many hundreds of pounds. This dangerously large mass has three major components: batteries, motors, and structure. Lead-acid batteries able to drive a mobile robot for a day contribute about one-third of the weight. But nickel-cadmium aircraft batteries weigh half as much, and newer lithium batteries can be half again as light. Electric motors are efficient and precisely controllable, but standard motors are heavy and require equally heavy reducing gears. Ultrastrong permanent magnets can halve the weight and generate high torque without gears. Robot structure has been primarily aluminum. Its weight contribution can be cut by a factor of four by substituting composite materials containing superstrength fibers of graphite, aramid, or the new material Spectra. These innovations could be combined to make a robot with roughly the size, weight, strength and endurance of a human.

The first generation robot will probably move on wheels. Legged robots have advantages on complicated terrain, but they consume too much power. A simple wheeled robot would be confined to areas of flat ground, but if each wheel has a controlled suspension with about a meter of travel, the robot could slowly lift its wheels as needed to negotiate rough ground and stairs.

The manipulation system will consist of two or more arms ending in dexterous manipulators. There are several designs in the research labs today, but the most elegant is probably that of the so-called Stanford Jet Propulsion Laboratory hand (mentioned above, now found at MIT), which has three fingers each with three controlled joints.

The robot's travels would be greatly aided if it could continuously pinpoint its location, perhaps by detecting the signals from a handful of small synchronized transmitters distributed in its environment. This approach is used in some terrestrial and satellite navigation surroundings, to find doors, detect obstacles, and track objects in its work space. Research laboratories, including my own, have experimented with techniques that do this with data from television cameras, scanning lasers, sonar transducers, infrared proximity sensors and contact sensors. A more precise sensory system will be needed to find particular work objects in clutter. The most successful methods to date start with three-dimensional data from special cameras and laser arrangements that directly measure distance as well as lateral position. The robot will thus probably contain a wide angle sensor for general spatial awareness, and a precise, narrow angle, three-dimensional imaging system to find particular objects it will grasp.

Research experience to date suggests that to navigate, visually locate objects, and plan and control arm motions, the first Volks-robots will require a billion operations per second of computer power. The 1980s have witnessed a number of well publicized fads that claim to be solutions to the artificial intelligence or robot control problem. Expert systems, the Prolog logical inference language, neural nets, fuzzy logic, and massive parallelism have all had their spot in the limelight. The common element that I note in these pronouncements is the sudden enthusiasm of groups of researchers experienced in some area of computer science for applying their methods to the robotics problems of perceiving and acting in the physical world. Invariably each approach produces by simple showcase demonstrations, then bogs down on real problems.

This pattern is no surprise to those with a background in the twenty-five year research robotics effort. Making a machine to see, hear, or act reliably in the raw physical world is much more difficult than naive intuition leads us to believe. The programs that work relatively successfully in these areas, in industrial vision systems, robot arm controllers, and speech understanders, for example, invariably use a variety of massive numerical computations involving statistics, vector algebra, analytic geometry, and other kinds of mathematics. These run effectively on conventional computers, and can be accelerated by array processors (widely available add-ons to conventional machines which rapidly perform operations on long streams of numbers) and by use of modest amounts of parallelism. The mind of the first generation Volks-robot will almost certainly reside in quite conventional computers, perhaps ten processors each, able to perform 200 million operations per second, helped out by a modest amount of specialized computing hardware that preprocesses the data from the laser eyes and other sensors, and that operates the lowest level of mobility and manipulation systems.

MIND CHILDREN (2050+)

The fourth robot generation and its successors, with human perceptual and motor abilities and superior reasoning powers, could replace human beings in every essential task. In principle, our society could continue to operate increasingly well without us, with machines running the companies and doing the research as well as performing the productive work. Since machines can be designed to work well in outer space, production could move to the greater resources of the solar system, leaving behind a nature preserve subsidized from space. Meek humans would inherit the earth, but rapidly evolving machines would expand into the rest of the universe.

This development can be viewed as a very natural one. Human beings have two forms of heredity, one the traditional biological kind, passed on strands of DNA, the other cultural, passed from mind to mind by example, language, books, and recently machines. At present the two are inextricably linked, but the cultural part is evolving very rapidly and gradually assuming functions once the province of our biology. In terms of information content, our cultural side is already by far the larger part of us. The fully intelligent robot marks the point where our cultural side can exist on its own, free of biological limits. Intelligent machines, which are evolving among us, learning our skills, sharing our goals, and being shaped by our vales, can be viewed as our children, the children of our minds. With them our biological heritage is not lost, it will be safely stored in libraries at least, but its importance will be greatly diminished.

What about life back on the preserve? For some of us the thought of being grandly upstaged by our artificial progeny will be disappointing, and life may seem pointless if we are fated to spend it staring stupidly at our ultra-intelligent progeny as they try to describe their ever more spectacular discoveries in baby-talk that we can understand. Is there any way individual humans might join the adventure?

ROBOTICS AND ARTIFICIAL INTELLIGENCE

You've just been wheeled into the operating room. A robot brain surgeon is in attendance, and a computer waits nearby. Your skull, but not your brain, is anesthetized. You are fully conscious. The robot surgeon opens your brain case and places a hand on the brain's surface. This unusual hand bristles with microscopic machinery, and a cable connects it to the computer at your side. Instruments in the hand scan the first few millimeters of brain surface. These measurements, and a comprehensive understanding of human neural architecture, allow the surgeon to write a program that models the behavior of the uppermost layer of the scanned brain tissue. This program is installed in a small portion of the waiting computer and activated. Electrodes in the hand supply the simulation with the appropriate inputs from your brain, and can inject signals from the simulation. You and the surgeon compare the signals it produces with the original ones. They flash by very fast, but any discrepancies are highlighted on a display screen. The surgeon fine-tunes the simulation until the correspondence is nearly perfect. As soon as you are satisfied, the simulation output is activated. The brain layer is now impotent—it receives inputs and reacts as before, but its output is ignored. Microscopic manipulators on the hand's surface excise this superfluous tissue and pass them to an aspirator, where they are drawn away.

The surgeon's hand sinks a fraction of a millimeter deeper into your brain, instantly compensating its measurements and signals for the changed position. The process is repeated for the next layer, and soon a second simulation resides in the computer, communicating with the first and with the remaining brain tissue. Layer after layer the brain is simulated, then excavated. Eventually your skull is empty, and the surgeon's hand rests deep in your brain stem. Though you have not lost consciousness, or even your train of thought, your mind has been removed from the brain and transferred to a machine. In a final, disorienting step the surgeon lifts its hand. Your suddenly abandoned body dies. For a moment you experience only quiet and dark.

Then, once again, you can open your eyes. Your perspective has shifted. The computer simulation has been disconnected from the cable leading to the surgeon's hand and reconnected to a shiny new body of the style, color, and material of your choice. Your metamorphosis is complete.

Your new mind has a control labeled "speed." It had been set at 1, to keep the simulations synchronized with the old brain, but now you change it to 10,000, allowing you to communicate, react, and think ten thousand times faster. You now seem to have hours to respond to situations that previously seemed instantaneous. You have time, during the fall of a dropped object, to research the advantages and disadvantages of trying to catch it, perhaps to solve its differential equations of motion. When your old biological friends speak with you, their sentences take hours—you have plenty of time to think about the conversations, but they try your patience. Boredom is a mental alarm that keeps you from wasting your time in profitless activity, but if it acts too soon or too aggressively it limits your attention span, and thus your intelligence. With help from the machines, you change your mind-program to retard the onset of boredom. Having done that, you will find yourself comfortably working on long problems with sidetracks upon sidetracks. In fact, your thoughts routinely become so involved that you need an increase in your memory. These are but the first of many changes. Soon your friends complain that you have become more like the machines than the biological human you once were. That's life.

THREE

BIOLOGICAL TECHNOLOGIES

Claude A. Villee

Our understanding of the concepts of molecular biology such as gene structure and the control of protein synthesis has greatly increased in the past decade. The technology to conduct experimental procedures unknown just 10 years ago, such as isolating and cloning genes, is also developing at a rapid pace. It is truly mind-boggling to imagine what the field of biotechnology might be like half a century from now if we look at the rapid progress over the past decade and the technologies now under development that will permit gene transfer to treat humans suffering from inherited diseases.

This article considers two apparently different but closely connected levels of biology—microbiology and macrobiology. Biotechnology works at the microbiological level to achieve effects that become manifest at the macroscopic level.

We shall break our study into two parts. In the first, we shall consider a number of subjects at the genetic and molecular levels. Under this, we shall see the structure and function of genes, with special attention to the features of genes occurring in eukaryotic (nucleated) cells. We shall also see the developing technology for repairing human genetic material that is defective, or altogether absent. We shall follow this with a discussion of the promise and possible roles of *catalytic antibodies,* which can be custom synthesized to perform specific enzymatic reactions. And finally, we shall examine the nature of *hybridomas,* an important class of hybrid cells and how they may be used in genetic engineering. In the second part we shall consider some of the macrobiological problems facing humanity today, including population growth, food supply, and human life span.

BIOTECHNOLOGY

The term biotechnology refers to the discipline that investigates, develops, applies, and commercializes the variety of techniques used in manipulating the molecules of life, as contained especially in genetic material and in proteins. Given the molecular similarities of viruses, bacteria, plants, animals, and humans, biotechnology has widespread implications in many areas of medicine, agriculture and industry.

The genetic material of living systems is located primarily in long double-stranded helical molecules called DNA (Deoxyribonucleic acid). These molecules are wrapped in segments around specific protein molecules and then compacted into a series of coils and folds to produce discrete packets called chromosomes. A DNA molecule is composed of units called nucleotides, and the two strands of the molecule are linked together by bonds between facing nucleotide units. Each nucleotide contains three molecular subunits. Two of these—a sugar, deoxyribose, and a phosphate group—are standard for all nucleotides. The third is a variable subunit (A,G, C, or T), which forms the bonds between the two DNA strands according to precise pairing rules: A pairs with T, and G pairs with C. The nucleotide combination A-T (adeninethymine) or the nucleotide combination G-C (guanine-cytosine) is termed a *base pair.*

Each chromosome is composed of many genes which are the basic unit of genetic material. Each gene contains the code for making one specific protein, and is a section of the DNA molecule consisting of thousands of nucleotides in a specific order. The sum total of the genes within all of the chromosomes in a cell is termed the genome. The cells of human beings contain 23 pairs of chromosomes and therefore the sum of all the genes in the 46 chromosomes constitute the human genome. More than a score of human genes have now been isolated, mapped, and sequenced. The complete sequence of nucleotides is known for these genes and their DNA can be synthesized.

The two complementary strands of a DNA molecule each serve a different purpose. One, the coding strand, serves as a template for the synthesis of many messenger RNA molecules (one messenger RNA for each of the many genes along the strand). Each messenger RNA in turn is used by the cell as a blueprint for making a particular protein molecule. Protein molecules are made of chains of molecular subunits called amino acids, which are strung together according to the RNA codes. Three successive nucleotide subunits of RNA (termed a codon) specify one particular amino acid. The second noncoding strand of DNA serves as a template for making copies of the coding strand.

THE GENOME PROJECT

There has been much interest in current proposals to determine the nucleotide sequence of the entire human genome. It is estimated that there are about 100,000 genes in each human cell, and each gene is composed of 30-50,000 nucleotides. Clearly, carrying out the sequencing project will require a very large investment of manpower, time, and money. This project would rank in scope and cost with the Manhattan Project that resulted in the production of the atomic bomb and more recently with the Superconducting Supercollider (SSC) that will accelerate elementary particles to very high energies.

DNA—The Molecular Book of Life

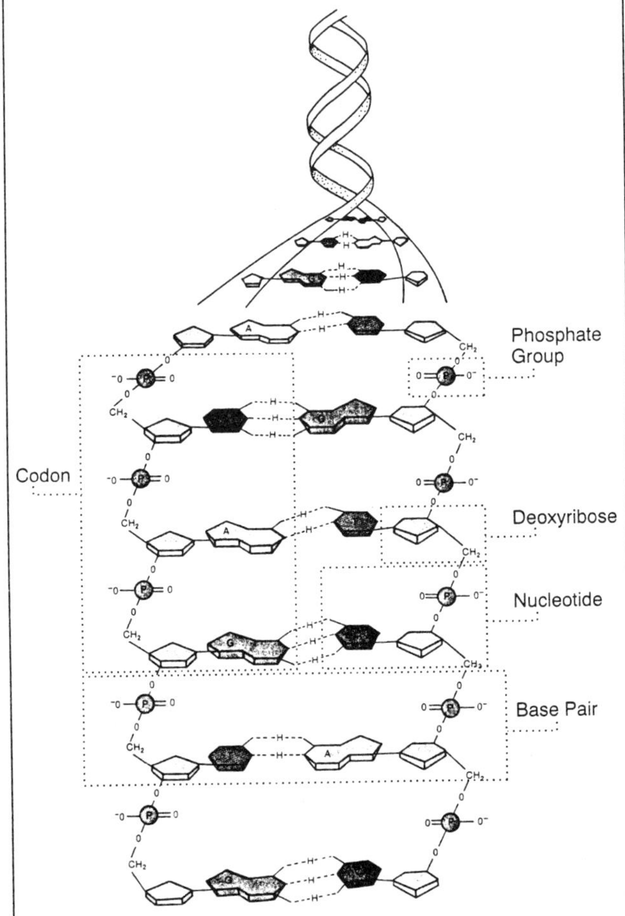

ADAPTED FROM VILLEE ET AL, *BIOLOGY*, 2ND EDITION, P. 304 (1989) SAUNDERS COLLEGE PUBLISHING, PHILADELPHIA.

One question that troubles many biologists and geneticists is: "What do we do with this sequence after we have determined it?" Proponents argue that this would provide us with valuable information that could lead to new ways of treating or preventing a great variety of inherited diseases. Opponents counter that the massive sums required for the human genome project will siphon funds away from other smaller but relevant projects that are more likely to yield understanding of particular medical problems. Whatever the conclusions of the present debates, it is quite likely that the genome map will be completed well before 2044, opening areas of research and technology that cannot now be foreseen.

GENETIC ENGINEERING

The procedures that involve the isolation, cloning, and sequencing of individual genes and the replacement of missing or faulty genes by intact genes from another source are often called genetic engineering.

In one of the great surprises of molecular biology, researchers discovered in the late 1970s that the original RNA transcript produced from DNA in eukaryotic cells contains extensive sequences of nucleotides that are not used in producing protein molecules. The prior concept had been that a single gene in the DNA molecule in the nucleus serves as the template for making an RNA molecule, which moves out of the nucleus and serves as a template for making a single protein molecule. Now an additional step has been added to the process. A single gene in the DNA serves as a template for making an original RNA molecule, which contains extra segments of RNA beyond what is needed to code for making one protein molecule. In the added step, these extra segments, called introns, are neatly excised, and the remnants, called exons, spliced together by specific enzymes called ligases. Then the resultant RNA molecule can leave the nucleus and serve as a blueprint for making a protein molecule.

The number of introns present in different genes may vary over a wide range. For example, the betaglobin gene, which determines one of the components of hemoglobin, contains two introns; the ovalbumin gene, which codes for one of the proteins in egg white, contains seven introns. The combined lengths of the introns may be considerably longer than the combined lengths of the protein-coding exon sequences. For example, the ovalbumin gene contains about 7,700 base pairs, whereas the sum of its exon sequences is only 1,859 pairs. Some of these introns have been conserved in the course of evolution and are found in the DNAs of a variety of organisms. Scientists believe the conserved introns must have some, perhaps important, role in gene regulation, but precisely what that role may be is unknown. One role discovered for the intron RNA is that it sometimes acts as an enzyme. This new class of biological catalysts has been termed *ribozymes.* Hopefully, long before the year 2044 the puzzle of what purpose is served by the introns will have been solved.

RECOMBINANT DNA

The decade of the eighties was marked by tremendous advances in the technology of recombinant DNA, which enables the molecular biologist to isolate a fragment of DNA containing a specific gene from one organism, transfer it to another organism, and replicate it within the host cell, usually a bacterium or a yeast cell. Millions of copies of the gene are produced as the host cell produces many generations of daughter cells. The production of a large number of identical pieces of DNA as the cell replicates is termed *cloning.*

The cloned DNA sequence can be modified or combined with parts of other genes to alter the genetic mechanism. By inserting the human gene that codes for insulin into a bacterial strain and growing the cells in culture it is possible to produce many copies of the gene and, subsequently, large amounts of human insulin to be used in the treatment of diabetes. A number of important pharmaceutical agents such as growth hormone and interleukin are being produced by comparable recombinant DNA techniques.

Many human diseases (more than 4,000 known so far) are caused by defective or absent genes, which cause defective or absent enzymes or other proteins. The possibility of permanently curing these diseases by identifying the gene that is missing, producing copies of that gene from other individuals, synthesizing the appropriate messenger RNA that codes for the enzyme, and then using the methods of revere transcription to produce genic DNA (using RNA as the template) that codes for the enzyme is very promising.

A very active field of research at present is that of devising methods for introducing the replacement gene into the DNA of the proper cells of an individual that lacks the gene, thus achieving a permanent cure. Getting the synthetic gene not only into the proper cell but into the nucleus of that cell is a difficult problem and several procedures have been tried. One method that is very promising at present is that of using specific viruses as *transport vehicles* to carry the genes into the cell. Certain viruses can readily penetrate certain kinds of cells and carry their genetic material into the cell's nucleus. There the replacement gene can be spliced with the cell's own DNA and become a permanent part of the genetic information of the cell.

Successful gene transfers have been achieved using several kinds of viruses, especially retroviruses (similar to the virus causing AIDS), which usually do not kill the cells they infect. Still under investigation is the nature of the necessary regulator elements that turn the system on and off under the proper circumstances. New strains of plants can be produced by gene transfer after the genes have been isolated, cloned, and inserted into plant cells using viruses as transport vehicles. Hybrid strains produced in this fashion may have increased resistance to infection, increased yields, or the ability to grow in salty soil or in soil irrigated with salty water. The solutions to these and other problems will make it possible, in the not too distant future, to replace the missing or defective gene that codes for the production of hemoglobin and thus provide a permanent cure for people who suffer from sickle cell anemia or from thalassemia.

The first federally approved attempt to replace a gene in a human was carried out in 1990, when a child with ADA (adenosine deaminase) deficiency was treated with cells containing the missing gene. Comparable studies with various genes that have been isolated and cloned will provide cures for other diseases such as cystic fibrosis and hypercholesterolemia. These methods will undoubtedly be perfected well before 2044 and will be an important method of treating individuals who suffer from the lack of the gene.

CATALYTIC ANTIBODIES

Antibodies are large proteins composed of light (low molecular mass) and heavy (high molecular mass) chains and of variable and constant regions of the protein chain. This complex structure permits the development of the large number of different kinds of antibodies. There are at least 100 million different antibody molecules possible!

One of the more recent findings relating to biotechnology has been the discovery that certain kinds of antibodies can be modified so that they acquire catalytic, enzymatic, activity. These catalytic antibodies can, at least in theory, be made specific for many of the thousands of enzymatic reactions that normally occur in living organisms. Several ways of generating catalytic antibodies have been developed. Each kind of antibody has structure that is complementary to the antigen, the protein used to generate the antibody. These techniques have been used to produce antibodies containing combining sites that complement the structure of the rate-limiting intermediate transition state of the reaction. Catalytic groups may be introduced into the combining site of an antibody by chemical modification. In this way antibodies have been produced, each of which will catalyze one of a wide variety of chemical reactions.

HYBRIDOMAS

Hybridomas are cells produced by the fusion of two other cells, such as spleen cells and myeloma (cancer) cells. The hybridoma technology permits the formation of large quantities of homogeneous antibodies. The host organism is immunized with a particular antigen, causing it to produce antibodies specifically directed against the antigen. The spleen cells that produce antibodies cannot be cultured in vitro (in a test tube) but are fused with myeloma (cancer) cells which do grow and divide very rapidly for an indefinite period of time. The hybridoma cells formed by the fusion of spleen cells and cancer cells can produce antibodies and can also be grown in cell cultures. The hybridoma cells are selected by growing the mixture of cells on a special growth medium. Unfused myeloma cells are unable to grow on this medium and are removed from the culture. The pure hybridomas are then cloned and separated into colonies. The colonies are then screened for their ability to generate antibodies that bind selectively to the original antigen and not to the carrier protein.

By appropriate chemical modification one region of the antibody can be converted into a form that has specific enzymatic activity. In this way it is possible to produce large amounts of an antibody with specific enzymatic activity. Specific enzymes can be designed and produced that will carry out specific tasks. Large amounts of the enzyme can be produced. For example, antibodies could be produced that would have the enzymatic activity needed to chew up oil that has been spilled into the environment and convert the heavier oils to lighter oils that would evaporate. Alternatively, a sequence of catalytic antibodies could be generated that would metabolize the hydrocarbons completely to carbon dioxide and water. Catalytic activity can also be selectively introduced into antibodies using the techniques of molecular biology.

Since the first reports of catalytic activities of antibodies in 1986 a considerable number of reactions have been catalyzed by antibodies. The specificities of these antibody-catalyzed reactions are quite high and the reaction rates have been accelerated as much as one millionfold. It is very likely that, in the coming decades, the technologies will be exploited to the full to produce catalytic antibodies that can be used effectively in solving medical, general biological, and ecological problems. The year 2044 will see attacks on these problems using very different kinds of tools from those used in the past.

POPULATION GROWTH AND CONTROL

One of the pressing global problems today is that of population growth. Indeed, it plays a central role in the development of many other global problems such as famine, disease and pollution. If population trends continue at their present rate the population of the earth in the year 2044 may well be more than ten billion, with most of the increase occurring in Africa, the Mideast, India, Southeast Asia and Latin America. In addition to the general tendency towards an increase in the total number of individuals in the world as a whole, there is a differential growth in the population of the cities as people in many nations move from the countryside to the cities, seeking better jobs and better living conditions.

Thus the year 2044 could see many megalopolises of 100 million or more people. This increased population density will have a marked effect on individual privacy and health, for much of the world will be a ghetto. Controlling population size is perhaps the most essential step to be taken by humankind in the coming years.

An effective method of control of population growth is by the use of contraceptives. It has been shown repeatedly that women of all races and ethnic groups will choose to use contraceptive techniques if they have ready access to ones that are safe, effective, and easy to use. Mechanical contraceptive techniques may be replaced by biological control methods based on our emerging understanding of the hormonal controls of reproduction.

Hormonal control of reproduction has been utilized widely in the past two decades but a critical issue is to determine what positive and negative side effects these "pills" have on the woman with prolonged use. Progesterone is a hormone that causes the lining of the uterus (womb) to grow and prepare for the nurturing of a fertilized egg. In the past few years an analog of progesterone that neutralizes its effects has been developed in French laboratories and is being used successfully in a few countries, notably France and China. This antiprogestin pill, RU486, is both safe and effective in creating conditions that prevent the implantation of a fertilized egg. The widespread use of RU486 on a worldwide basis could have a significant effect in reducing the rate of population growth. However, the acceptance of this method in certain cultures may be opposed on possible moral grounds. Another method (Norplant) recently approved for use in the United States is that of implanting tiny plastic tubes containing progesterone under the woman's skin. The progesterone gradually diffuses out of the tubes, into the blood stream and into the tissues where it produces its contraceptive effects. The tubes need to be replaced every five years or so.

Another contraceptive method now being developed is the synthesis of analogs to the gonadotropin releasing hormone, GnRH (secreted by the hypothalamus, a part of the brain, which would bind to receptors for luteinizing hormone (LH) in the pituitary gland. This would prevent the LH surge that normally triggers ovulation, the release of an egg from the ovary. These analogs are small peptides and they possibly could be designed to be taken intranasally rather than by injection. Interfering with ovulation using GnRH might be more acceptable than interfering with implantation using the RU486 pill.

Interfering with the hormonal control of the reproductive process may become the most effective method for reducing population growth in the next decade. The problem of population growth has already been solved in certain regions of the world such as Europe and in certain majority populations of the U.S. and the C.I.S. Much more needs to be done in other parts of the world and other cultures. This requires that every couple in the world has access to the information and the means to carry out effective contraceptive practices. It is a reasonable expectation that by the year 2044 the use of reliable, safe, and inexpensive methods of population control will be widespread in all parts of the globe and population growth may well be controlled at last.

FOOD SUPPLIES

A doubling of the size of the human population would require at least a doubling of the food supply—possibly even more to remedy some obvious imbalances in the availability of food to different regions and population groups. Since the world's population is expected to increase another billion, to a total of over six billion in the next decade (and most of the increase will occur in Third World countries that have the lowest food supplies), there is great need for increasing the world's food supply. It is estimated that some 1.6 billion humans are malnourished at the present time. Where subsistence farmers have great difficulty in eking out an existence on land depleted of its nutrients, new agricultural techniques are needed. Unfortunately, current world food production has practically reached a plateau and has begun to fall behind the growth of the population. There is much concern at present that the oceans have been overfished and we are in danger of having greatly reduced supplies of food from this source.

Some 10 million hectares of agricultural land are degraded and become unproductive annually and must be replaced. An additional five million hectares must be converted to agricultural uses to supply food for the increased number of humans. Essentially all of the land converted to agriculture is obtained by deforestation.

The present methods of intensive agriculture can provide some increase in the amount of foodstuffs raised around the world but there are limiting factors such as water supplies, soil depletion, and the increasingly large amounts of energy required per unit of output. Some 35 percent of the total food produced is destroyed by a variety of pests, such as insects and rodents, despite the use of pesticides. Without the use of pesticides the destruction of foods would be considerably greater and food shortages would be even greater.

Efforts to increase food production by conventional methods would require a vast increase in the amount of energy devoted to agriculture (fertilizers, pesticides, mechanization, etc.) and would lead to increased pollution of both air and water and increase the loss of precious topsoil. A great deal of energy is lost in the food chain as herbivores such as beef cattle eat the primary producer, the grasses, and humans eat the beef. The earth could feed more humans if all the cattle and sheep were killed and humans ate the grasses and other vegetables directly. Eating tenderloin steak may be a luxury in monetary terms, but eating hamburger is equally luxurious in ecological terms.

It is for that reason that biotechnology can make an important contribution to food production. For example, increases may be achieved in part by mapping the genome of the major crop plants—wheat, corn, and rice—that will allow the development of transgenic strains with combined high yields and pest resistance.

The methods of increasing food production by the techniques of molecular biology are not restricted to state-of-the-art laboratories in scientifically advanced countries.

Farmers and scientists in scientifically less developed countries are finding that with technical assistance they can carry out biotechnology processes using simple apparatus for growing plant cells in culture flasks. This technique allows the farmer to improve the quality of plant varieties that are locally available. Some of the first results were the growing of potato plants from cells in culture. The cells from special strains of potatoes developed at the International Potato Center in Peru have been grown in culture to yield mature potato plants with improved crop yields and increased resistance to pests. Ongoing experiments have shown that the tissue culture techniques can be applied to a variety of other vegetables and fruits, raising the possibility of increasing production of edible plants. This is particularly important for areas where access to good quality seed stock has been a limiting factor for agriculture production.

LIFE SPAN

One factor contributing to the growth in population is that advances in medical and public health techniques have led to an increase in the average human life span. As a result, by 2044 a greater percentage of the population will reach the age of 100 before dying. However the ultimate length of life of individuals will probably remain at about 125 years, because the process of aging of cells seems to proceed inexorably by mechanisms not clearly understood at present. This is an important area of research in the future. Some of the solutions may be obtained by using biotechnology measures. Recent clinical trials suggest that administering growth hormone made by the methods of recombinant DNA to elderly patients increases the protein content of their bodies and decreases their fat content. This gives them the appearance of being more youthful and may prolong their life span. An additional challenge posed by the increase in the life span is to assure an adequate quality of life for the increasing elderly population without placing an excessive burden on society.

Maintaining the quality of life for the elderly is another important area for biomedical research.

THE GREENHOUSE EFFECT AND GLOBAL WARMING

One of the widely discussed possible threats to the future of mankind is that of global warming due to an increase in the carbon dioxide content of the atmosphere over the past century. This increase can be attributed to the burning of fossil fuels such as coal and oil and the stripping of forests and the burning of the trees to obtain energy. This results in the conversion of the hydrocarbons and carbohydrates to carbon dioxide and water. It is postulated that carbon dioxide and other gases such as water vapor and methane accumulate in the atmosphere, forming a layer that reflects downward the thermal radiation rising from the earth. This raises the temperature of the globe just as though it were a gigantic greenhouse. Extensive reviews of temperatures in many parts of the world suggest that the average global temperature may have increased some 0.5°C over the past 100 years. The increase is far from uniform, with some areas in the world having temperature increases and others decreases. In the past century there was a period of worldwide cooling from 1950-1970. Records of the concentration of carbon dioxide in the atmosphere have shown a small, gradual increase over the past century. Computer programming of possible climatic changes that might be brought about by the accumulation of carbon dioxide in the atmosphere has led to the identification of the so-called "greenhouse effect."

The reality of the greenhouse effect is still controversial among experts. However, if there is such an effect due to carbon dioxide there are several measures that can be taken to minimize or prevent the accumulation of the carbon dioxide layer. This would require a substantial decrease in the amounts of coal, oil, and gas that are burned and a great increase in the use of solar and nuclear power, which do not contribute carbon dioxide and water vapor to the atmosphere.

The amount of carbon dioxide produced by the burning of gasoline in automobiles could be reduced if new means of mass transportation were designed and built. The use of biomass fuels such as ethanol or methanol derived from corn and other plants would also contribute to the production of carbon dioxide. On the other hand, the production of biomass utilized carbon dioxide from the atmosphere, so that the production and utilization of carbon dioxide are in rough balance. Perhaps some totally new fuels can be developed using the methods of biotechnology. The dynamics of the use of carbon dioxide in the earth's crust and atmosphere are not yet well understood.

The carbon dioxide in the atmosphere is utilized by green plants to synthesize carbohydrates, and a further strategy for minimizing the greenhouse effect is to plant trees and shrubs. Some fast-growing trees and shrubs can grow well even in poor soils. A large faction of the carbon dioxide in the atmosphere is taken up by microscopic green plants (algae) in the oceans and lakes. Any measure that would increase the uptake of carbon dioxide by these microscopic plants could also have a marked effect in stabilizing the amount of carbon dioxide in the atmosphere. One factor that may limit the growth of the algae and other plants that live in the ocean, particularly those living in the southern hemisphere, appears to be the amount of iron in the water. Other possible factors limiting the growth of algae, for example, the amounts of nitrogen and phosphorus, may be supplied by "acid rain" which contains nitric and other acids. It has been suggested that providing the oceans with additional iron might lead to a great increase in the fixation of carbon dioxide by the microscopic plants present, thus decreasing the amount of carbon dioxide in the atmosphere. However, the amount of carbon dioxide taken up by the algae probably could not balance the amount of carbon dioxide resulting from the burning of fossil fuels.

The amount of carbon dioxide produced by the burning of gasoline in automobiles could be reduced if efficient electric cars could be produced and if people would use public transport systems. By 2044 there may be relatively little petroleum remaining to be used in public transport.

The initial prediction about the rise in water levels in the ocean due to the melting of the polar ice caps caused by the global warming range from 0.60m to 7m. More recent ones suggest that the increase in water level will be no more than a few centimeters.

CONCLUSION

To solve some problems threatening the future of mankind using genetic engineering, new techniques must be developed to extend the plethora of techniques already invented and in use. This will require more research and financial support. Some of these techniques and procedures have been attacked on ethical grounds. Perhaps the ethical arguments will have been resolved in the next half-century and appropriate guidelines for further advances in biotechnology will have been established. The major factors limiting biotechnology are scientific and technical.

Biotechnology is among the most intensive and active areas of research at present and promises to continue at a rapid rate over the forthcoming decades. Research will certainly continue in the areas discussed here, but we can only guess as to what totally new findings will appear and have major effects in shaping the subject in the future. What new avenues of research will open up and what effects, positive and negative, will these findings have on human life in the future?

FOUR

BIOMEDICAL TECHNOLOGIES

Ernest G. Cravalho

A few years ago there appeared on television two shows, "The Six Million Dollar Man" and "The Bionic Woman," that portrayed the adventures of two individuals, one male and one female, who were the amalgams of normal humans and the best that fictionalized biomedical engineering had to offer. Not surprisingly, these television shows precipitated in many of the young people who watched them a desire to become the biomedical engineers who could create such marvels. As a result, the numbers of people entering the profession of biomedical engineering have grown significantly in recent years.

In the coming decades, as these young engineers live out their fantasies, we should expect to see The Six Million Dollar Man and The Bionic Woman become realities. If life truly imitates art, by the year 2044, humans will be fitted with a variety of electronic and mechanical devices that will enable them, at the very least, to carry on "normal" lives (if not perform superhuman feats) when they otherwise might have suffered lives of incapacitation due to trauma and disease. While these achievements are noteworthy in themselves, advances in biomedical technology over the next 50 years will extend way beyond these examples.

Precisely what these accomplishments might be is difficult to predict at this time. In looking back at the advances in biomedical technology since 1944, it would appear at first glance that no one could have predicted from that vantage point where we could be in 1994. In fact, the growth of biomedical technology has been so great during the past decade alone that one could not have predicted its progress 10 years ago. However, while we might not have envisioned the details of the great strides in areas such as magnetic resonance imaging (MRI) technology during the last decade, we could argue that the general directions in which biomedical technology was heading over the last 50 years were obvious in 1944.

Historically, there have been some technological themes common to every decade as long as there have been technologists and practitioners of health care. These themes are just as likely to be relevant in the half century between now and 2044. Innovators of biomedical technology have been motivated in large part by the need to aid the physician. Over the past century, medicine has been practiced largely on the basis of observation: The physician makes certain observations about the physiological condition of the patient, and based upon this knowledge and experience from treating patients who evidenced similar conditions, the physician prescribes a treatment.

The success of the treatment thus depends upon the physician's knowledge, experience, quality of observations, and the efficacy of available therapies. To enhance the probability for success in this scenario, it would be necessary to improve the quality of the physician's observations and of the therapies available.

SCIENCE IN SUPPORT OF MEDICINE

In the earliest stages of practice of the healing arts, the physician had to rely upon the senses of sight, hearing, touch, and smell to determine the physiological state of the patient.

As experience showed that the senses, in their natural form, were insufficient to the task, effort was devoted to extending the limits of these senses. The stethoscope (invented in 1837) and the X-ray machine (invented in 1895) are examples. Over the past century, a significant portion of the activity in technological advancement in medicine has followed this theme. The computed tomography (CT) scan and magnetic resonance imaging (MRI) system are outstanding examples of the way in which this theme has continued to the present day. CT technology uses computers to reconstruct a three-dimensional image from many minute, precision-controlled X-rays. MRI produces three-dimensional images of internal structures based on signals released from protons that have been stimulated by an intense magnetic field.

Barring a major revolution in the practice of medicine (although even if physicians were replaced by robots, the thesis would still be true), we can expect significant effort to be devoted to the innovation of new diagnostic technologies in the next 50 years.

The forms that these innovations will take will be shaped by many factors, not the least of which will be economic. The public is going to be more concerned about the costs of new technologies and about who will pay for them. People will want hard answers to questions of the need for new and expensive diagnostic technologies. Cost/benefit ratios will become more prominent in setting policy governing technological development. "Do we really need this new diagnostic device? Can we do without it? Does it add any truly new information?" These are questions that will be asked more and more in the years ahead.

In addition, the impetus for these new advances is not going to originate with the physician, but with the technologist. As technologies become more complex, the technological breakthroughs will be based upon details known only to one who specializes in the art. The invention of the CT scan by physicists and mathematicians and the development of MRI technology by physicists and chemists are evidence that this shift has already begun.

In the case of therapies, the situation will be somewhat different. The developments here will be a result of the mutual efforts of physicians, technologists, and life scientists. Therapeutic technologies cover such a spectrum of needs that their innovation cannot be focused in any one sector. However, it is likely that much of the effort in therapeutic innovation will be devoted to the development of systems that will replace the function of organs, tissues, or even impaired cells. Some of these functions may be replaced by hardware and software emanating from the creative genius of technologists while others will be more biological in origin and will derive from the innovation of biologists, chemists, physicists, and physicians.

Here the public is going to demand answers to tough ethical and legal questions: "Is it morally right to do this? What is the social impact of this new technology? Who must suffer to gain this advance? Is the quality of life truly improved by this new technology?"

BEYOND THE HUMAN EAR

The stethoscope is closely identified with the physician. It is the first technological device that the young physician learns to use for diagnostic purposes and with which he can literally work diagnostic miracles. But the stethoscope is sorely limited in its ability to collect acoustic signals. Furthermore, the signals are processed by the physician, whose own transducers (ears) may be imperfect.

As a result, the wealth of acoustic signals emitted, transmitted, absorbed, and reflected by the tissues of the patient are not fully exploited. While it might be argued that the range of long wavelengths and sensitivity associated with the acoustic signals processed by the human ear do not give sufficient resolution to be helpful in locating pathologies with any precision, it is true that we have not fully exploited the potential of the technology. First of all, the technology need not be limited to the acoustical spectrum of the human ear. Second, the technology need not be limited to the "sounds" emitted by the patient. Third, the processing of the acoustic signals need not be the physician's responsibility alone. And finally, the behavior of the signals need not be limited by the acoustic characteristics of the tissue of the patient alone.

Diagnostic ultrasound is an extension of the stethoscope that incorporates the four points above. It uses the sounds emitted by a transducer outside the patient, and the reflected signals are processed with algorithms that convert them into a visual record that lets the physician "picture" the organ being studied. To aid this process, the ultrasound technician often makes use of lenses or coupling devices that provide better access by modifying the acoustical characteristics of the tissue. But this is only a single example of what can be done.

In the future, the physician will utilize a "stethoscope" that collects signals over a broad frequency spectrum and then analyzes this spectrum for information content with the aid of sophisticated mathematical techniques programmed into a microprocessor. The recorded information will then be displayed in a form quite different from the sounds heard through the conventional stethoscope. The process can also be complemented by introducing sounds via devices similar to endoscopes placed inside the patient. The processing of the information will use mathematical models describing the acoustic behavior of normal and diseased tissue. In the ultimate application, it is possible that a patient could be placed in an anechoic (sound free) chamber and "scanned" acoustically. Since the radiation employed is nonionizing and hence less harmful to tissue, it offers obvious advantages over conventional X-ray technology.

ACTIVE CAPSULES MONITOR BODY CONDITIONS

New devices in the form of active capsules equipped with sensors, transistors, and batteries will be developed so that information of a biochemical nature can be collected (acoustically and otherwise) as these capsules travel through the patient's anatomy after being swallowed. Information then will be transmitted to a receiver outside the body where it will be processed and displayed on-line. The physician will thus have access to a patient's interior without resorting to invasive means. Taken sequentially, these devices can show the time history of the interior physiology. Advances in microfabrication technology will facilitate manufacture of these devices at low enough cost to make them truly disposable.

Current imaging technologies already provide visual information of almost sufficient spatial resolution so that the diagnostician has a complete visual picture of the patient's anatomy, and this will be extended further in the years ahead. But the existing systems, particularly MRI, are not fast enough. That is, the "throughput" (the number of records processed per unit time) is presently limited by the speed of the electronics creating the electromagnetic fields necessary to collect the data. As the speed of the electronics increases in the coming decades, it will be possible to increase the throughput. Moreover, with the aid of the new high-speed electronics and computers, imaging systems will provide a more complete "picture" of the patient, including biochemical data, in real time. Thus, the diagnostician will have a complete cinematographic visual record of the anatomy and the biochemistry of the region of interest that shows the way in which this "picture" changes in real time.

There will be so much information the diagnostician will be unable to keep up with it all. For this reason, some of the functions of the diagnostician will be assumed by "expert systems" using artificial intelligence. This computer-based approach will complement the diagnosticians work, reducing the time required for diagnosis. In true artificial intelligence form, these expert systems will be capable of "learning" as they progress through a diagnosis so that in the future a similar diagnosis can be realized in a fraction of the time.

The use of expert systems will not be confined to the processing of visual records alone. In the intensive care setting, similar expert systems will be coupled to patients via miniature transducers capable of monitoring any physiological parameters. The data thus collected will then be used to predict the course of the patient's pathology and to recommend appropriate courses of action so that care can be optimized. Again, the expert system will serve to extend the physician's ability to process vast amounts of data in short time intervals. The ability of the expert systems to learn in this context will be invaluable in responding to the needs of the critically ill in a timely fashion.

Other diagnostic technologies will exploit the potential of the laser as a compact source of monochromatic, coherent light that can be directed to any site in the body through fiber optics. The new diagnostic technology will employ exogenous chromophore (chemical compounds that become optically active when irradiated with light of the right wavelength). They are exogenous in that they are not naturally found in the body. They can be targeted for specific sites by coupling them to monoclonal antibodies that will bind to specific receptors in the body. Thus the presence of specific pathologies at specific locations can be detected photo-optically thereby providing considerable spatial precision and specificity. The information so obtained can be both quantitative and qualitative in nature. In the coming decades, this diagnostic technology will be routinely available for virtually any pathology and will be used to complement diagnoses available by other means, such as MRI and CT.

THERAPEUTIC TECHNOLOGIES

As significant as these diagnostic technologies may be, they will be useless unless appropriate technologies for therapy are developed as well. There will be a need for significant advances not only in pharmacology but also in the means of delivering these drugs.

In the next 50 years, drug delivery systems capable of delivering the appropriate dose at the right place at the right time under active control (either bioelectronically or by the central nervous system) will become commonplace. These systems in many cases will be biodegradable in the sense that the implantable packaging of the drug will breakdown biochemically after use. This technology offers one of the most effective means for managing patient care while also minimizing the risks of drug overdose. The ability to localize the treatment, which is inherent in the design of these systems, is absolutely essential in many therapies.

In cases where surgical intervention is necessary, minimally invasive techniques will reduce patient morbidity and mortality by reducing the need for anesthesia significantly. In addition, costs will be substantially reduced due to shorter hospital stays. Advances in this area will be accelerated by advances in laser technology. Since laser energy can readily be directed to any site in the body, it will be an invaluable component of minimally invasive surgery. By properly selecting pulse duration, wavelength, intensity, beam width, and other laser parameters, it is possible to use the laser to cut, weld, seal ablate, and coagulate biological tissue. The process will be automated by exploiting the laser's diagnostic capabilities under control of an expert system that can process this information and then direct the therapeutic operation of the laser.

In the coming decades, the manufacturing industry will be revolutionized by the use of robots and robotic techniques. Much of this technology will spill over into medicine, so that robotic prosthesis *á la* The Six Million Dollar Man will become routine with control provided by the central nervous system via the neural network of the body. However, even more significant will be the coupling of microfabrication technology with robotics, resulting in the development of microrobots which will course through the body on a "Fantastic Voyage" via the circulatory, respiratory, and other fluid systems. These microrobots under external control will be able to collect physiological data in a fashion similar to the capsule diagnoses mentioned earlier, but more importantly, they will be able to perform therapeutic tasks such as microsurgery and drug delivery.

THE MOTHER OF HER GRANDFATHER'S SISTER

Two therapeutic technologies likely to see significant development in the next 50 years merit special mention because they will raise ethical and legal questions of enormous proportion. These are transplantation technologies and reproductive technologies. These two technologies have already become linked in a single case in 1990, in which a child was conceived for the specific purpose of providing a bone marrow transplant to a sibling. In the future this situation will become further exacerbated as the technology of cryopreservation advances further and enables the entire reproductive process to be carried out *in vitro* employing reproductive materials (sperm and egg) that have been cryogenically preserved. The resulting embryo can then be cryopreserved for generations so that it is possible, with the aid of embryo transfer, for a woman to give birth to her grandfather's sister.

Of greater concern is the use of this technology to provide spare parts. For example, by manipulation of a two-cell embryo or by splitting (using a laser) an embryo more advanced in development, two genetically identical twins can be created. One can be transplanted while the other is frozen for future use as a source of "spare parts" for subsequent transplantation. Clearly, the ethical implications of this technology are overwhelming.

No less important are the ethics of transplantation. During the next 40 years the immunological issues associated with xenografts, that is transplantation of cells, tissues, or organs from one species to another, such as mouse to human, will be solved. Thus it will be possible to cure diabetes by transplanting islet cells (the pancreatic cells that produce insulin), from a mouse, or some other animal, into a human. This could lead to the raising of animals for the sole purpose of providing xenograft. Some might question whether one species has the right to use another in this manner.

This problem will be further exacerbated by the development of biohybrid, artificial organs, in which a bioreactor is charged with cells from an animal organ—for example, a rat liver, which is then used in an extra-corporeal fashion to detoxify the blood of a human whose own liver was damaged by disease or trauma. Clearly, this technology would be of enormous benefit to liver patients, but it does raise questions about the need for controls to safeguard the exploitation of other species.

Finally, it is likely that in the next 50 years technology will enable humans to access nerve impulses and to decode these signals for the purpose of controlling artificial limbs. Of course, the next step is to transmit signals that would control natural limbs so that the motion of human subjects can be controlled remotely. Thus the Cyborg becomes a reality. Extending the technology further, it should be possible to control the operation of the central nervous system so that mind control becomes a reality. Will humans then become automatons under the control of some clever biomedical engineer? While the danger may be real, it is not imminent, so there is ample time to develop the appropriate safeguards.

Biomedical technology offers the very real possibility of extending life significantly with even the very real possibility of improving the quality of life. Its potential raises serious ethical questions that are just as worthy of our attention in the coming decades as are the technological issues from which they are derived. Just as all technological problems are solvable in the eyes of the technologists, so too are the ethical issues in the eyes of a rational society. But the two must be addressed simultaneously if we as a race are to survive and prosper.

FIVE

TRANSPORTATION AND COMMUNICATION

S. Fred Singer

Transportation and communication are basic human activities today and fulfill basic human needs. Their rapid expansion and diversification in the twentieth century have been the most visible of the changes that have taken place during this time. How will emerging technologies further develop and affect transportation and communication in the next century? It is logical to first discuss future needs (the demand) for transportation and communication services before delving into the supply side, which is conditioned by new transportation technology.

FUTURE DEMAND

Can we expect significant breakthroughs in transportation and communication—comparable to what has happened in this century—besides the ongoing spread of existing technologies? It is this question that we want to address in this essay.

It is convenient to divide the discussion into three parts: the transportation of goods, the transportation of people, and the conveyance of information. Historically, it is likely that the transportation of goods preceded the transportation of people. In former times, people did not travel for leisure as we do today, nor did they have to commute to work. (Whole populations migrating in search of food or in the herding of animals, etc., cannot be considered under transportation as we understand it.) Similarly, the conveyance of detailed information over long distances and short time spans had to await at least the invention of writing.

TRANSPORTATION OF GOODS

In this regard, for reasons of clarity, we shall maintain a necessary distinction between *point-to-point* transportation, which involves movement of goods in large bulk, and *distribution,* which entails transporting goods in relatively smaller quantities from one or more central points to a larger number of outlets.

As far as point-to point transportation is concerned, the present modes would seem to be adequate. For example, petroleum constitutes one of the most important bulk materials in world trade, and can be transported at lowest cost by pipelines over land, barges over rivers and canals, and tankers over oceans. (Pipelines are the only means that permit continuous transportation; other modes must transport materials in batches.)

Other raw materials of commerce—such as coal, lumber, sand, gravel, and such—are also transported in bulk. Traditional modes of overland transportation are fairly cost efficient and should be adequate for as long as these materials continue to be used in the future. High-value cargo, however, can command air transportation because the faster travel time reduces in-transit inventory costs.

The rise of cities has brought with it the problem of distribution of finished products and foodstuffs to consumers. This task is quite different from the point-to-point transport we have discussed up until now, or the multipoint transport of raw materials and semi-finished goods into factories. It is the distribution phase of transport that involves the greatest per unit cost—and by the same token lends itself best to the application of new technology designed to reduce this cost.

The cost of distribution is greatly increased by traffic congestion. Indeed, congestion may set a limit to the economic size of a city and define what constitutes an optimum city size. In practice, political factors allow cities to grow well beyond optimum size—and by virtue of the greater population acquire even more political clout to grow further and more uneconomic. Ideally, optimum-sized cities, spaced apart for low-cost point-to-point transport, would provide the most economic means to house a population.

TRANSPORTATION OF PEOPLE

Let us consider the transportation of people and its future in productive activities, in the services sector, and for leisure and recreation.

Productive activities, such as agriculture, the extractive industries, and manufacturing, require workers—and these workers have to commute to and from the work site, usually on a daily basis.

It is likely that in the future, general technological advances—not transportation technology per se, but robotics—will make a considerable difference by requiring fewer workers to achieve the same output, thereby reducing the need for transportation. (The classic example is agriculture, where the fraction of population involved in the United States was about 80 percent two centuries ago and is now less than 5 percent—and the United States exports food as well.) The trend toward automation is likely to accelerate as sensors, control techniques, and computers develop further and become lower in cost.

As opposed to productive activities, the services sector is increasing rapidly in size; but the number of people involved may actually decrease in the future, especially where information exchange is involved. Again, advanced technologies can replace people. Within the next decades video conferencing may replace meetings. Telerobotics and the "electronic office," with the use of facsimile and other sophisticated telephonic services may reduce the need for travel even further. And then there are a variety of technologies that are moving along, replacing people and thereby the need for transportation: electronic banking comes to mind, perhaps electronic editing of newspapers and journals, and other similar activities.

It is unlikely that electronic systems will take over completely. There is something very human about meeting with people: it fulfills an important emotional need. Certainly, medical examinations of sick people will not be conducted entirely by telemetry and video techniques; the human touch is important in more ways than one. But one can visualize an increasing degree of replacement, and therefore a lessened need for transportation of people.

The matter of psychology and sociology is even more relevant when we discuss recreation and leisure travel, which today is the fastest growing sector of transportation. Here there are important questions. Will communication substitute for transportation? Will grandparents be willing to see their grandchildren over a video screen rather than touch them in person? Will TV and more advanced audio-visual technologies, such as TV holography, become substitutes for travel—as the National Geographic magazine used to be not so long ago? Will people be willing to experience Bali or Alaska vicariously? Or will they prefer the expensive, time-consuming, and not always pleasurable process of visiting these places? Is it likely that tourism will continue to grow, as per capita incomes increase around the world—providing more leisure time and more discretionary spending? The answers to these questions are of obvious importance for tourism, airlines, hotels, and other affiliates of the travel industry who depend for their survival on this multibillion dollar and growing market.

COMMUNICATION DEMAND

In many ways predicting the future of communication is simple because we have already seen much of the future. Electronic transmission of information is rapidly replacing the traditional modes of communication. Facsimile and inter-modem transmissions are fast replacing letters. Although these modes are much faster, they are, however, less personal. Will books and newspapers still be printed and transported 50 years from now?

So far, the availability of new and efficient modes of information transfer is keeping up with the rapidly increasing volume of information. Optical fibers are replacing copper wires, widening the information channels. Information storage devices are becoming cheaper as magnetic and optical storage technologies battle it out. Information handling is becoming faster and cheaper as new processes are being developed. There is little doubt, therefore, that the technology "in the pipeline" will be sufficient to handle the information explosion in the next decades. The real constraint on growth in the foreseeable future, however, may be the absorptive capacity of the human mind itself.

The existence of these technologies, in turn, will impact on recreation and leisure travel. The nature and extent of that impact, however, is difficult to predict. It does seem reasonable that as films, theater performances, and travel experiences are further converted to more advanced electronic means, more people will prefer to receive these experiences in their homes through their enhanced-realism television receivers, rather than venture out.

NEW TECHNOLOGIES

Inevitably, future technologies will affect the demand for, as well as the means of, transportation and communication. But it is the latter that concerns us in what follows.

It would seem that for transportation the greatest contribution will be in raising convenience and speed, with convenience certainly being the primary consideration. At the same time, technology will decrease the cost and increase safety.

Space travel and exploration should be the big adventure of the future. Currently, space travel is confined mainly to astronauts and conducted primarily in the areas of communication, earth and weather surveillance, and even manufacturing. In the future, space travel may largely be engaged in recreation and leisure, that is, for the sheer adventure of leaving the parent planet and being able to view and contemplate it from a distance.

Modest extensions of current technology could make space travel for the public quite feasible. It would seem to be limited only by its costs and the willingness of people to pay. However, I am not sure that commercial travel will ever use the ballistic-missile route through space—even though it would shorten the New York-Tokyo trip to about an hour.

A different development relates to communications. A widely discussed innovation is the "personal newspaper," which does not require any new technology and can bring you exactly the kind of information that matches your interests. It allows you to get all of the headlines and then dig into any stories that you wish to explore. It may even allow you to interact with the writer on a real or time-delayed basis; it can provide you with hard copy when needed and allow you to comment on it as you wish.

The communication scheme will quite probably be part of the larger scheme of the "electronic office," allowing you to stay in touch with your work, friends, and the news and entertainment that you want. Again, current technology is capable of handling these developments.

But in spite of such electronic means of communications, transportation will still be required. Turning first to personal transportation, let us focus on the automobile. It is difficult to imagine that anything will replace, in the next 50 years or so, the convenience and speed of the automobile. It will survive in its essential form because it is so useful, but it will acquire some new features. Improvements will be made that will reduce cost and the weight of the materials used, improve energy efficiency, and reduce, if not altogether eliminate, the pollution caused by exhausts. Considering that automobiles currently kill about 50,000 people per year in the United States alone, representing a huge human and financial loss, safety will be a major consideration of all technological advances.

As far as an automobile's efficiency is concerned, two points can be made: 1) internal combustion engines can be designed to run very efficiently and with very little pollution if their speed is constant; and 2) most of the energy in city traffic is used in acceleration and wasted in deceleration. Taken together this suggests a future—perhaps the early twenty-first century—in which electric cars, running on low cost, low weight batteries, will take over. These batteries will be charged by small internal combustion engines running at constant speed, and therefore high efficiency. And they will use some kind of "dynamic braking," so that the kinetic energy of the car can be converted in electric or mechanical energy on deceleration, stored, and reused when the car accelerates.

Another of the new features will be "smart cars" and perhaps "smart highways." Smart cars involve methods of navigation and information that allow them to travel more efficiently, avoiding congestion. Smart highways will allow cars to travel automatically, following guide wires embedded in the pavement. We can already see the beginning of smart cars today, as many acquire on-board navigation systems, and simple traffic control experiments are being conducted in different parts of the world. With respect to smart highways the wait may be much longer, simply because of the cost. The technology is within reach; but the cost of modifying highways is high and as such it is not worthwhile until enough cars can take advantage of the system. Conversely, car owners are unlikely to invest in the system until a major fraction of highways is so equipped.

Also far in the future—certainly many decades away—is the personal rapid transit (PRT) system. Although many studies have been done on it, its cost would seem to put it beyond immediate reach. In its most elegant form, you can visualize stepping into a small cabin, holding perhaps two people, and pressing some buttons that designate the destination. Thereupon, the vehicle is automatically guided, using rail or embedded wire, to arrive at your destination in the shortest possible time. The concept can be further expanded by letting the PRT cabin become part of the "modular expanded airplane," invented by Federal Aviation Administration scientists Machol and Lupinetti.

A less costly approach can be used to reduce congestion and pollution, even within the next decade or so. The best way to do this would be to impose fees for both congestion and pollution. Congestion can be measured quite easily and the fee adjusted accordingly. Collection of this fee might be a problem. One would not want to place a tollbooth on every street corner. But technological means are now becoming available that can recognize cars and bill them automatically as they pass a given location—the so-called "smart tollbooth." No personnel would be required, nor would the car have to stop. Similarly, remote sensing techniques are becoming available that can measure the pollution in the exhaust of cars as they pass by, photograph the license plate or recognize the car by some automatic means, and bill a pollution fee or provide other means of rectifying the situation. By properly adjusting congestion fees and pollution fees, almost any desired policy can be achieved. For example, it should be possible to direct traffic so that cars take advantage of low traffic densities, and to move ownership of polluting cars into regions of the country away from urban areas, where the residual pollution of the vehicle does not impact significantly on ambient air quality and where pollution taxes could, therefore, be lower.

Urban mass transportation is a perennial subject of great interest. Contrary to popular opinion, fixed rail is quite expensive, not energy efficient, and in many instances requires large public subsidies. Bus systems are less expensive because they use the existing highway infrastructure. But the great objection to mass transportation is its lack of convenience. Fixed-rail means point-to-point transportation, and most people require area-to-area transportation. They need to get from their home to the mass transportation terminal either by bus or by car, and likewise they need to get to their place of work at the other end. One way of solving this problem is by a hybrid transport system involving mass transit plus personal transportation. The personal transportation consists of small personal vehicles, holding perhaps two people—perhaps electrically driven, like a golf cart—that can be rented at a mass transit terminal for a nominal sum, but need not be returned to the same terminal. The role of technology would be to lower the cost of the vehicle sufficiently so that it becomes a practical scheme.

In the meantime, taxicabs, and especially jitney cabs, can provide many of the advantages of mass transportation and personal transportation—without waiting for new technology. We see it, for example, at air terminals, where a limousine will transport a group of people to a certain area and then deliver them to individual addresses. Again, there are no technological barriers, but lowering the cost would certainly make this kind of service much more attractive and widespread.

One of the chief characteristics of transportation throughout human history has been the increasing speed of transportation—from walking and running at a few miles per hour to supersonic transports (SST) traveling at twice the speed of sound. There is great popular fascination with speed, and a tendency to measure advances in transportation by the maximum speed capability of the vehicle. For example, in high-speed surface transport there is magnetic levitation ("maglev"). Two different approaches are being used, one in Germany and one in Japan, using different methods of magnetically levitating a train so as to eliminate friction and move the train at speeds approaching 320 kilometers per hour (200MPH).

So far at least, maglev has not fulfilled its promise. One reason is that the cost of the track is very high. In the meantime, the extension of conventional steel-wheel-on-rail techniques—using better methods of preparing rails and roadbeds, avoiding curves and grades, and controlling the motion of railcars running on these rails—have produced extremely high speeds. In Japan and in France interurban trains reach speeds in excess of 240 KPH (150MPH).

But for longer distances, and for traffic that does not have high density, one can find no better solution than air transport. This is the reason that commercial aviation has captured not only long-distance passenger transportation, but has even beat out railroads and trucks for freight transport.

Air transport technology is moving in two different directions. One is to service smaller communities that cannot afford large airports. Here the emphasis is on aircraft that can land and take off from short runways. STOL (Short Take Off and Landing) are the designation used.

But the "holy grail" of air transportation seems to be supersonic, and even hypersonic transport (HST)—aircraft traveling at speeds faster than three times the speed of sound. But the HST is certainly not suited for passenger travel, and even an extension of the existing SST, the Anglo-French Concord, may be in doubt. On a planet the size of Earth, it makes little sense to travel at speeds higher than Mach 3: It takes too long to reach top speed and too long to slow down to landing speed. Studies so far indicate that supersonic speeds of about Mach 2.5 are optimum for long-distance air travel. Even so, the costs are very high unless new technologies for materials and engines can come to the fore.

On the other hand, it may make sense to transport freight supersonically over long distances, particularly if one can construct an unmanned supersonic freighter that can travel without frills, without crew, and under automatic control, delivering freight back and forth on a routine and continual basis. Some economic studies indicate that, subject to the development of a jet engine and reliable remote control systems, such a freight-carrying SST may be economically more feasible than a passenger SST, and available in a decade or two.

Government has traditionally played a strong role in transportation and communication—in many countries even owning the facilities, including national airlines, railways, and telephone and postal systems. The history of government in the development of transportation and communication technology is a mixed one, and its future role is ambiguous. New technology often requires a helping hand from government, particularly in regard to environment and safety, but in other cases innovation comes from private entrepreneurs. Yet transportation and communication are so vital to modern society and have so many ramifications, both good and bad, that an assessment of all of the impacts of new technology is required. Planning a future in transportation therefore involves a partnership of the private sector and of governments.

SIX

INHABITING THE OCEANS

Athelstan Spilhaus

To the choice of where to live and work on land will be added the choice of living and working at sea. Within the next fifty years we will colonize the oceans and occupy them in the sense of going to sea to stay, living there, working there, and coming back to land only for exotic vacations.

Man speeds up everything. Evolution in nature is slow and thus it is very difficult to predict where it will go. But evolution of man-made devices goes so fast that it is possible to extrapolate into the future. Such extrapolation should be exponential. Linear extrapolation into time provides predictions that are generally outstripped by reality. The reason the extrapolation must be exponential is because the number of minds working on new choices increases with the number of minds on earth—with population—which increases approximately exponentially.

Future cities at sea and using the ocean as real estate for people's living are merely the logical evolution of things that have gone before. For centuries, the Melanesians and Polynesians lived on atolls which, even though not mobile, provided a completely maritime existence. These people achieved some of the greatest feats of navigation among their isolated ocean-bound homes, driven by their need to expand their bartering, their fishing, and the circle of their friends.

The Indians of Bolivia living on Lake Titicaca spend their lives on the water on artificial islands constructed of totora reeds. In the Orient, many generations of river people eked out their living on sampans which they never left. These people had no *pied-á-terre* so they made their own *pied-á-mer*. Singapore, one of the most densely populated and best run cities in the world, is essentially a sea city.

Lighthouse keepers, although stationed on promontories and rocks, really led their lives at sea. Man-made evolution toward life on or near water continued from lighthouses to lightships and, during World War II, when crossing the Atlantic was a hop-hop proposition, to weather ships that reported conditions along air routes. There was talk of floating airfields at sea, but rapid and spectacular advances in aeronautics made these unnecessary.

NEED

Necessity and invention are akin to the chicken and the egg. Necessity breeds invention but invention, in its turn, breeds further necessity. It is the need for space for a growing population that will make the colonization of the oceans a certainty as well as a need. At the same time, the advances in marine technology, not only of the size of ships but also of their amenities, provide the means for colonizing the three-quarters of the planet Earth covered by water, which is now "waste space."

The need is not only physical—for space—but instinctive—an inborn desire to live near water. Water spawned all life on earth. The creatures that were our ancestors crawled from the seas onto the bits of land that were the result of tectonic churning; but those creatures that survive on land only do so if there is water. Land is useless without water.

The sea is in our veins and in our primeval memories, so it is no wonder that people flock to rivers, lakes, inland seas, and ocean edges until the shorelines are crowded. The newborn child is soothed by the rocking of the cradle because, in our early evolution, we rocked with the waves.

All the great teeming cities of the world owe their locations to water for the support of commerce. Their dense populations not only ravish the land but also overuse, overgraze, and overpollute the precious water that gives the city people life. As the limited shoreline becomes too crowded, people will move out onto the water!

In the United States, there is approximately four times the density of population along the shorelines as there is in the interior. In Canada, the ratio is about five to one, and in Australia it is even higher. This attraction to the sea is true worldwide, but the shoreline is finite. It has been estimated that, even with California's long stretches of beaches, if everyone in California went to the beach on the same day, they would have to stand shoulder to shoulder and more than ten people deep.

While human beings love to be near water, let us remember that they are the worst polluters on earth and, when crowded together, are themselves a pollutant. Their discarded substances endanger not only their own lives but also other life on earth, particularly marine life along the shorelines and in the estuaries. After all, the shorelines and estuaries, which are nurseries for fish hatchling, are among the most fragile regions on earth. Just as people protect the nursery of a loved child, they must protect the shorelines—especially from themselves.

The policy of all nations should be to encourage, even offer incentives for, people to move away from the shorelines. Programs are needed to foster the migration of populations inland—or out to sea, as described shortly.

SPACE AVAILABLE

The space available for homes, recreation, and industrial sites is almost unlimited at sea. Already, countries have generally agreed upon extending their territory into the oceans for a distance of 200 miles. This is where the first ocean cities will be built. Locations in what are now international waters will follow soon afterward.

Japan, where population pressures on very limited land makes moving out to sea urgent, is already building cities, airfields, power plants, industrial sites, and recreational parks on man-made islands near its shores.

However, man-made islands too close to the shore still may interfere with the estuaries and shorelines of the land, the cradle of much of the life in the sea. Many endeavors can be better pursued in the open ocean further away from the areas that are vital to the preservation of coastal waters. For instance, it would be better to continue oil pipelines out to sea so that tankers need not come into an ecologically fragile shore, but would tie up at a floating platform where both tanker and platform rise and fall together with the level of the sea.

PURPOSE

To get further away from the shorelines, floating cities will be built to carry out enterprises—ocean enterprises. Almost anything one can do on land, one can do at sea; but the purpose of floating cities is to provide real estate for those enterprises that can only be done at sea or that can be better done at sea.

Let us consider what we can do best at sea. Sea cities are excellent sites for ocean bottom mining and fishing support, ocean farming, and marine protein processing. They are ideal for energy production, either by oil drilling in the seabed, ocean thermal energy conversion, or wave energy conversion. Ocean industry can also utilize, at sea, without the costly transmission to land, the energy to accomplish such things as the desalinization of seawater and waste treatment and conversion. Sea cities are ideal oceanographic research platforms and harbors; radar and weather sites; military bases; space vehicle launch platforms; and assembly plants, packing plants, and processing plants. They are also suitable for retail trade (free ports); commercial airports; prime oceanfront recreation centers, resorts, conference centers, and even theme parks—Disneyseas to counterpoint Disneylands! The city can be designed to accommodate many specialized purposes.

One enterprise that can be carried on in ocean cities at a profit is to take care of the refuse from the crowded land cities. High temperature incineration in the open ocean with low residual ash has been proven to be environmentally acceptable. This will bring land money to the ocean cities so that the ocean cities can import from land those things which are best grown or made on land. Just as cities on land grow their own crops and import supplemental foods from the sea, so sea cities will grow their own seafoods and import supplemental foods from land.

Ocean cities will combine many enterprises and purposes in a coordinated effort so that the "platform," or floating island, could become economically viable. Although no one purpose might justify such expensive real estate, the many profitable interlocking ocean enterprises make it economical.

Cities on land grow unplanned. Self-contained ocean cities will be built for fixed occupancy and will handle all their problems as in a well-designed ship. They will not lend themselves to uncontrolled growth as on land.

Ocean cities provide an opportunity for a step that all politicians on land have dodged, namely, the control of population. A precondition for living in ocean cities must be limited occupancy. We promote this on land only in elevators and transportation vehicles. On ships at sea, the danger of overloading has been recognized from earliest times. Thus, ocean cities may lead to the recognition that planet Earth itself can only bear a limited occupancy.

BUILDING

How will a floating ocean city be built? Like anything else, by building modules and assembling them piece by piece. Just as houses are built brick by brick so will ocean cities, the only difference being that the bricks in this case are much bigger. But it is still essentially modular fabrication and assembly.

In the simplest case, the city would be constructed of floating, steel reinforced, boxlike modules which would be hinged and winched together to provide as large a configuration as desired.

A "basic" brick or module could have a surface area of 100 meters square and be fabricated in a dry dock. Twelve of these modules might be assembled in quiet water into a rectangular aggregate or section 200 meters by 600 meters; later on, they might be fabricated at sea. If an airstrip were required, three aggregates could be towed to the location in the open ocean and there be linked together into a 200 meter by 1,800 meter strip.

Aggregates could be added as requirements dictate, including planned increases of population without violating fixed occupancy per aggregate, and assembled in whatever configuration is suitable for a particular enterprise. For instance, an area encircled by modules could form a large and deep enclosure of stilled water suitable for the breeding and domestication of sea mammals, even whales.

The tools, instruments, ship designs, materials, power plants, and the foundation of marine science are at our fingertips. Of course, there are engineering challenges such as the stability of a huge floating artificial island in the face of the winds against the city, and the currents on the parts of the city beneath the sea.

Just as construction on land has an upward trend, creating increasingly taller building, so construction in the open ocean will be downward. By dipping down in the cold water, the space within the floats that keep the city's head above water and hold the platform steady can be utilized fully. The floats go down deep enough to make use of the cold water at this depth. Refrigeration and air conditioning might occupy the bottom of the floats, industry and clean manufacturing the middle, and living space for people, the very top.

GOVERNMENT, LAW, AND REGULATION

The first steps toward the colonization of the seas will be cities supported by the seabed, where water is shallow, or floating in the deeper territorial waters of nations that border the sea. These will be governed by the legislatures of the nations in whose waters they grow. From the start, they will have their own mayors, judges, police, and emergency brigades.

Soon after, sea cities, essentially floating islands, will move into international waters. A floating city with power for navigation is truly a ship. Under international law, it can fly its nation's flag. It is guaranteed free passage, even if only traveling at a very few knots, as long as it keeps moving, even under tow.

When the numbers of people living at sea are sufficient to form a constituency, landlubber bureaucrats will be forced to come down to earth and, realizing that earth is mostly sea, respond to the voice of the sea people. There will be representation of the sea people in legislative assemblies—congressmen and senators from the State of the Sea.

Inevitably, individual sea cities will enter into treaties with others, form federations, and seek independence. Elections will be held, and there will come about government of the sea people, by the sea people, and for the sea people.

SEVEN

LIVING IN SPACE

Gerald K. O'Neill
(The late, deceased 1992)

People move to new frontiers for better economic opportunity, to escape political or religious oppression, or to live in better or more exciting surroundings. All three reasons, but especially the first, are likely to motivate the early settlers of space. Their jobs will be in an industry that must be located in space itself to be economically viable. There may be only one such industry initially, but it could become so large as to require a substantial population of highly educated workers in space together with their families. That industry would build solar power satellites for supplying clean energy to Earth.

Space has strong advantages over the surface of moons or of other planets as the site for location of human habitats (colonies). In space we can build habitats that are large in scale—far larger than could be built on a planet, because those structures will not have to withstand gravity. But they can rotate, to provide for their inhabitants internal gravity equal to Earth's norm, for which we have evolved over millennia. A space habitat will be a spherical shell of metal and glass, enclosing a normal, breathable atmosphere—something that does not exist on any planet other than Earth.

In space, clear of the shadow of a moon or planet, solar energy can be used fulltime for every energy need, from climate control to the growing of crops, grass, trees, and decorative flowers. Like outposts on the surfaces of the Moon, or Mars, space habitats will need protection against cosmic rays. That will be obtained by enclosing each rotating, spherical habitat in an outer shell, not in contact with the rotating structure. The shell will be made up of silicates, the waste slag from the industries in space that will use lunar materials. It will be thick enough to shield the interior from cosmic rays as effectively as Earth's atmosphere and magnetic field do.

Within the strengths of ordinary materials, space habitats could be very large indeed—several kilometers in diameter. But the early ones will be much smaller, as little as a hundred meters across, and will be designed for a few hundred inhabitants.

Among the many reasons for extending the human ecological range to habitats in free space rather than to the surface of an existing body is the potential land area available in the long-term. There is enough "leftover" material in the asteroid belt to build space colonies with a total land area several thousand times that of Earth.

To understand the reasons for which colonies will be built in space, we must be guided by logic and practicality. We start with the world as it is today, and focus on the compelling worldwide needs. For our view ahead we depend on technologies that already exist. But we still cannot predict a time scale. New technical progress may speed the breakout, but cannot be planned for. And technological practicality is only one necessary ingredient for progress. Humans can accomplish great things in incredibly short times when their motivations are high and managerial environments are right, but their ability can be wasted interminably with nothing to show for it when warfare, bureaucracy, religious hysteria, or personal greed and selfishness dominate the decision-making process.

The global realities of today include a population expanding rapidly in the poorest nations; an increasing need for energy to enable these nations to make the industrial transition to comparative affluence and, as a result, low population growth; increasing atmospheric pollution, due mainly to the burning of fossil fuels; rapid changes in eastern Europe caused by the sudden collapse of the Soviet economies; the possession by many nations of stockpiles of atomic weapons, many of which are clandestine, and the continuation of conventional warfare by small nations, keeping Africa, Asia, and parts of Europe, Central and South America in poverty and anguish. Good signs include the astonishingly rapid industrial transitions to affluence made by those nations of the Pacific Rim that are not at war and are practicing market economies.

The chief global physical problems are intertwined: how to obtain enough energy, and how to reverse the destruction of the environment. United Nations projections indicate that the world of 2044 must use energy at five times the present rate, if poor nations are to rise out of poverty. Obtaining that energy from fossil fuels would very likely complete the destruction of the biosphere. Obtaining it from nuclear power would require the simultaneous operation of 63,000 large (1,000 megawatt) nuclear power plants. They would have to be spread through every nation, including those with weak educational and managerial infrastructures. The specter of more Chernobyls makes nuclear power an unpopular choice.[1]

Ocean thermal power, and geothermal power from the tapping of heat in the earth's magma, both suffer from the disadvantage of low efficiency. They would dump great amounts of unwanted waste heat into the biosphere as the price of the energy they might provide. Solar power is not available at night and under cloud cover, and its low conversion efficiency makes it, too, a source of large amounts of waste heat.

The most promising way to generate the amounts of energy needed by the world of 2044 appears to be the Solar

Power Satellite (SPS.) The SPS concept was invented by Dr. Peter Glaser in the late 1960s. It consists of collecting solar energy with large arrays in 24-hour orbit, where intense sunshine is available nearly around the clock. The energy is converted to radio waves, which are sent to large, fenced-off collecting antennas on Earth. There the radio waves are converted to electric power for the power grid. The intensity of the radio waves (at a frequency of 2.4 gigahertz) is far less than that of sunlight, so birds would not be hurt by them. Because the efficiency of conversion at the collecting antennas is more than 90 percent, very little heat is wasted: less than a tenth as much as in the cases of coal or nuclear power.

The technologies for an SPS are well understood and have been tested thoroughly. They go back to the radar of 50 years ago. One hurdle remains: How to place the SPS that would supply to the earth electric power equal to the output of ten nuclear plants.

There is a straightforward way to solve the lift problem: obtain the materials for the SPS arrays from the surface of the Moon, rather than lifting them out of Earth's tight gravitational grip. Fifteen years of research, supported mainly by the Space Studies Institute in Princeton and partly by NASA, have brought each of the essential technical buildingblocks for materials transport and processing to the successful proof-of-concept state. The launch vehicles now in use are adequate to the task of lifting the relatively small, modular procession and manufacturing units needed to seed an SPS construction industry in space.

Space habitats and solar power stations do not require any new scientific developments. Engineering and design of specific systems is required. The successful technology demonstrations of lunar processing and transport, and research on closed-cycle agriculture show that we can use space resources here on earth. It is in that context that we can outline a plausible future leading to the habitation of space. We image it as the following "future history."

In the decade leading to the millennium year 2000 the wealthiest nations, released from the 40-year threat posed by the Cold War, focused increasingly on economic issues. Competition between the European, Asian, and North American economic communities made and broke industries, and became major election issues. Well-educated labor available at low cost in the newly democratic countries of eastern Europe began to compete as Korean and Taiwanese labor had competed in earlier decades. Environmental concerns, previously stifled in the communist regions, spread worldwide as the skies darkened with pollution and Cold-War barriers to communications fell. By the year 2000 the rising tensions between the rival economic blocks made the prevention of new major wars a serious political issue. International cooperation on large-scale projects that would benefit all nations appeared as an attractive, vote-getting alternative to violent conflict.

Growing recognition of the need for clean energy combined with growing fear of the degradation of the biosphere. The multinational communications consortia, Intelsat and Inmarsat, by then several decades old, were taken as models for a new consortium to develop Solar Power by Satellite. By 2005 the SPS consortium was formed, and many nations had signed a treaty that called for the renunciation of fossil fuels when satellite power was readily available to all. The teeth in the treaty were economic: a boycott of products made in countries that continued to burn coal and oil.

By the year 2044 the industry in space, processing lunar materials into satellite power stations, was building more than 400 new stations every year, with a revenue value of more than six trillion dollars. Although the industry in space was highly automated, people were still needed: by 2044, more than half a million of them. Fifty space colonies, each housing 10,000 people, orbited Earth about two-thirds as far away as the

Moon. They sparkled in the night sky like a circlet of jewels shining in the sun.

A visit to one of them takes a day of travel. We approach from the side, and see the bright reflection of sunlight off the glass of the agricultural areas, which are like giant bicycle wheels set next to each other. With the habitation sphere, hidden within its shield of silicates, the agricultural wheels rotate about an axis that points always toward the sun. External mirrors reflect the sunlight through windows into the habitation sphere and the glass-roofed agricultural wheels.

Our spaceship, built to shuttle between low Earth orbit and the colonies, moves carefully to a position on the axis of the colony's rotation, and begins to rotate once every 30 seconds, matching the colony. We feel our ship nudge the docking port. The latches close, and we float through the opened hatch to the zero gravity corridor of the colony's spaceport.

After drifting through the wide corridor for a few hundred meters we emerge into bright sunshine. We are some 300 meters high on a hillside that rises steeply to our feet. A river winds through the bottom of the valley, and beyond the river another hillside rises to our own height. The landscape is Mediterranean: groves of olive trees are below us, on the terraced hillside, and far down the slope we see vineyard. There are flowers everywhere, and steep mountain paths descend from our position on switchbacks toward the valley floor. People are walking, others jogging, on the paths below us, and on the valley floor, 400 meters away, we can distinguish bicyclists.

Of the 10,000 inhabitants of the colony, more than a thousand are children. On the near and the far hillsides we see houses and small apartment buildings, none more than two stories high. Their roofs are of red tile, and their walls have the warm, rich earth colors of the old towns in Italy and the south of France. Their windows and their terraces are large, because the colony enjoys a mild climate year-round.

That was set by the colonists themselves, who chose their hours of daylight by scheduling when the ever-present external sunlight would be admitted.

To left and right our view could not be found on Earth. Both valley and river rise on either side of our field of view, joining above us to complete a circle. The river is, in fact, close to the "equator" of the colony, and we have entered close to one of its poles. We look upward to what is for us the top of that circle, and find the water surface above us facing down. A pair of canoeists is paddling, hundreds of meters away and above, and we see that their heads point toward what we call "down."

Before us there is an open train, a small version of those that carry visitors across large parking lots on Earth. We board it with our baggage, which weighs almost nothing, close to the pole of the colony. The train carries us slowly at an angle down the mountainside, and as we descend our weight gradually returns. At the valley floor our weight is normal, and our sensations are those of people in an especially pleasant spot on Earth. We have arrived, and our next days will be spent exploring this oasis in space, filled with light and life.

Note

1. The editor believes that these estimates by the author are high. In the first place, the UN energy projections for the year 2044 are, in the editor's opinion, exaggerated, but even if the factor of five proves to be correct, the total global rate of energy consumption would only be about 40 TW and not 63 TW, as the author indicates. In addition, not all primary energy will be electricity. Even if the world becomes highly electrified, electricity would not be more than 70-80% of the primary energy, that is about 30 TW. Some of that electricity will be generated using hydro and solar installations. Even neglecting these sources, if all the electricity were to be generated in nuclear plants, the number of 1000 MW nuclear plants needed would be 30,000 and not 63,000 as the author claims. However, most probably the total energy consumption by 2044 will be much less, not exceeding 30 TW, taking into account improvements in efficiency and conservation. Therefore, the nuclear power contribution should not exceed 15 TW, requiring less than 15,000

nuclear plants (compared with about 500 at present), spread all over the world. It is reasonable to expect that by 2044 most countries will have the educational and managerial infrastructure to operate them. And since no more RBMK-type nuclear reactors are being, or will be built, no more Chernobyls are possible. Thus the world in 2044 will have to depend more upon nuclear power than at present to satisfy our ever-growing demand for energy without inflicting irreparable damage on the environment. That does not mean that we should not try to use all available "benign" energy sources, including the SPS system proposed by the author. —*Marcelo Alonso*

PART TWO

FROM TECHNOLOGY TO SCENARIO

EIGHT

THE BIOLOGICAL CENTURY

Gregory Benford

The nineteenth century was dominated by the metaphors and technological implications of two sciences: chemistry and mechanics.

To be sure, the audacious Darwin-Wallace theory of evolution by natural selection began preparing the ground for modern biology, and excited enormous public furor. Elsewhere in England, Michael Faraday and James Clerk Maxwell were laying the foundations of electromagnetic technology. But the older crafts and models of Newtonian mechanics and workaday chemistry drove the great economic and social engines.

Yet in the waning years of that century, Edison and others sounded the opening theme of the next era. For clearly, physics has dominated our century.

Electromagnetic theory and experiment gave us the telephone, radio, television, and computer, and made the internal combustion engine practical, thus the car and airplane, leading inevitably to the rocket and outer space. The fateful wedding of that rocket with the other monumental product of physics, the nuclear bomb, led to the end of large-scale strategic warfare—as profound a change as any in modern times.

Even now, as the century wanes, physicists remain our scientific Brahmins. They dominate government committees, holding forth on topics far beyond their nominal expertise—defense, environmental riddles, social policy. Yet in our growing environmental problems and the rapid advances in other laboratories, far from the physics departments of the great campuses, a clarion call is sounding through our time.

I believe that we stand on the threshold of the Biological Century. While the particle physicists desperately try to get their Superconducting Super Collider built in Texas, against growing opposition to the $9 billion price tag, a smaller initiative proceeds: the Human Genome Project. This vast effort, eventually costing about $3 billion, will map the human genetic code—our DNA.

The Project's director is James Watson, co-discoverer of DNA with Francis Crick. It is the largest job ever attempted in biology, but surely not the last foray of biologists into Big Science, where physicists are already right at home.

I recently heard Watson talk about the ethical implications of being able to know who has defective genes. "What will it mean when we can be sure we're not all born equal?" he asked, and I had to admit it was probably going to scare a lot of people. Insurance companies will not want to cover people with a genetic predisposition to illness, for example.

But those are short-term ethical questions, surely. The true solution lies in fixing genes, not merely reading them. If parents-to-be can have their problem genes edited into normal ones, most of the issues may evaporate.

And this is just one of many advances which portend much. Will we stop at cleaning up what we see as defects? I doubt it.

As we all saw in grade school, once you learn how to read a book, somebody is going to want to write one—that's how authors are made. Once we know how to read our own genetic code, someone is going to want to rewrite that 'text' tinker with traits—play God, some would say.

But that lies a few decades off, I believe. The first signs of a quiet revolution in our daily lives will probably come with some pretty noncontroversial commercial products. Much research has gone into cellular critters which can digest oil spills or other toxic contaminants. Some work reasonably well already. Soon enough such research will give us a spectrum of organisms which digest unpleasant substances. That should mean refineries which don't stink, rivers that don't catch on fire and aren't sewers.

Plants have plenty of chemical defenses, and a smart farmer will come to use that. In temperate zones, winter is the best insecticide, it keeps the bugs in check. The tropics enjoy no such respite, so plants there have developed a wide range of alkaloids which kill off nosy insects and animals. Nicotine is an excellent insect foe, the fact that we addict to it is a curious side effect. Adapting such defenses to orchards and crops is an obvious path for batik.

Consider the farm of the next century, which we might better call a "pharm"—because it may well be devoted to growing proteins, not wheat. Already researchers can synthesize proteins in animals by co-opting their own schemes for making, for example, milk.

Genetically altered goats have been made to yield in their milk a particular *human protein* which effectively dissolves the fibrin clots responsible for coronary occlusions. Efficiencies are low, but probably won't remain so. To get high yields, it will be a good idea to go to the dairy cow, which produces 10,000 liters of milk a year.

Imagine a Guernsey which yields insulin, the expensive aid to diabetics. We could make such a cow by editing its genes which control the cow's internal chemistry. The simple way would be to make two kinds of Guernseys, one which produces milk rich in the "alpha" chain that helps make up insulin, and the second which makes the "beta" chain. This would free the cows from having to contend with insulin in their own systems, for only when the alpha and beta chains are mixed do we get insulin itself.

Insulin grown down on the pharm would probably be much cheaper than ours today. Similarly, there seems no barrier to making many pharmaceuticals in natural systems. Sheep might be specialized to a whole range of useful drugs, for example.

Sheep, goats and cows would become the essential "bioreactors," reproducing themselves in a barnyard batik which could benefit many farmers who never heard of protein tinkering. But there will be troubles, because such animals don't breed true. A dairyman in Argentina will have to come back to Pharms Unlimited for his next calf. Indeed, Pharms Unlimited would be made to make its cows so they can reproduce their patented! technology without a fat fee. So the Third World may see this as just another way to keep them on an unending economic string.

Such technology will spread into the immensely profitable realm of direct consumer goods. Let's start with some simple items, kinds so commonplace they can be advertised on television.

Imagine a kitchen cleanser which dissolves waste in those hard-to-get places, maybe even invading the grouting of tile in pursuit of fungus. Or a bath mat which slowly tugs itself across the floor, slurping up puddles, deposits of soap and hair spray, hairs, general 'human dander'. It lives on the stuff, plus an occasional helpful dollop of diet supplements from the otherwise distracted homemaker—who thinks of it as a rug, not a pet.

But many products, the opening wedge, will be less startling—in fact, that's why they'll come first. Resident 'toothpaste' that does the essential policing up after lunch, and maybe even makes your breath smell, well, not so bad. Stomach guardians which ward off Montezuma's Revenge before you notice a single symptom—permanently, because the microbes are symbiotic with you, and live throughout your digestive system.

Much further along, let's open up our imaginations. Maybe there will be a fashion in biocorduroy, which lives off your sloughed-off-skin, perspiration—and even, if you like, some of your less agreeable excretions.

The theme here is biological balance—what's waste to one creature can become food to another—with a desirable job done in the transaction. This is "homeostasis," the biological equivalent of the thermostat.

Tougher jobs, such as mining, could be done by 'worms' which forage, digest and concentrate desirable elements. 'Oysters' could lie on the sea bed, amassing from the water a valuable 'sand' the biotechnicians like—gold, platinum, even mercury.

Of course, agriculture is the ancient batik. For many millennia we've been breeding cows and corn, collard greens and collies, all to our whim. We can expect more exotic foods, of course, but more important, we may see new and better ways of growing them.

Ponder a subtle marriage of the acacia ant with the orange tree. In the wild, the ant serves the acacia tree by eating nearby saplings and competing weeds (a weed, after all, is not a biological category, it's merely a plant somebody doesn't like), by fighting off invading insects, and other functions a plant can't handle.

In return the ant gets to eat delicious parts of the acacia, which have evolved as just such an enticing reward. Suppose we generalize that scheme to our orange crops, or any other plant we want to protect. Away with insecticides! And the farmer doesn't have to spend nearly so much time checking up.

One of the troubles with such apparently open-ended future projections is that we have no firm idea what the limitations on batik will be. Chances are, they'll be wilder than we think (imagine the Frenchmen who rode hot air balloons gazing at the lunar crescent, and trying to glimpse the century-long path that led through the airplane and the rocket to Tranquility Base).

From our blinkered perspective, a Biological Century looks like a fundamental shift in worldview. Once we learn the trick of reproduction *a la* nature, not *a la* factory, we may see a collision between the classical economy of scarcity and one of bioplenty. Thinkers like Freeman Dyson have been pointing out that the specters which haunt our present—strip mining and burning up our dwindling resources—may be as narrow a vision as was Spain's obsession with taking gold out of the New World, while missing tobacco, the potato, "love apples" (tomatoes) and the rest.

Batik opens the promise that the truly fundamental resources will be sunlight, water and land—privileging the tropical South and 'green tech'. This could neatly turn the tables on the industrial, 'gray tech' North which will develop the batik in the first place. (Spain sent Columbus, but missed the boat conceptually in the following century.)

An immense payoff for a small, but self-reproducing investment of 'smart' batik is a daunting possibility. We primates dropped jackrabbits into Australia before we knew their long-range impact. This point is not lost on the new Luddites of our time, the Jeremy Rifkin crowd which fears any batik product, and considers animal husbandry as "slavery."

Perhaps such potential plenty can stave off the obvious problem faced by the "Third World"—already an antiquated term, for there is no 'Second World' left, beyond a few wilting dictatorships. Batik crops—the Green Revolution—have already helped the starvation problems in the tropics. But I suspect batik alone cannot stop the rapid growth of human numbers there. So what will happen?

The right term might be 'die-backs'—sudden, catastrophic collapse of whole life support structures on a regional level, the Four Horsemen writ large. I believe, though, that two social forces will bring even more dire events.

Consider: we will have a North with many accomplished bioengineers. Given our desire to extend our own lifespans, much research will have gone into an intricate fathoming of the human immune system, to fixing our cardiovascular plumbing, and the like.

On the other hand, the North will increasingly be appalled with the South's runaway numbers. Mega-cities will sprawl, teeming with seedy, corrupt masses. Torrents of illegal immigration will pour over borders. Responding to deprivation, crazed politico/religious movements will froth and foment, few of them appetizing as seen from a Northern distance.

The more the North thinks of humanity as a malignancy, the more we will unconsciously long for disasters.

Somewhere, sometime, someone will see a simple solution combining these two forces: the Designer Plague. An airborne form of, say, a super-influenza. The Flu From Hell, carried on a cough, with a several-week incubation period, so the plague path will be hard to follow. Maybe fine-tuned, too, carrying a specific trait that confines it to tropical climes, as malaria is (mostly).

Would anyone be mad enough to kill billions, hoping to stave off the ecological and cultural collapse of societies? It seems despicable, mad—and quite plausible, to me. Speculations along these lines have already been voiced by Mark Martin, a molecular biologist. The zealot could well come from citadels of high moral purpose, too. After all, the great mass murderer of our century came from the culture of Mozart and Goethe, and was a vegetarian.

Such dark possibilities come with any major advance in human capabilities. Only by anticipating them (H.G. Wells foresaw atomic war) can we do the thinking and imagining that might prevent them.

THE WORLD OF 2044

Science fiction has seen big changes coming before, and done a reasonable job at avoiding our getting blindsided. We did a workman-like job on the rocket, the bomb, the robot and computer. Still, it's worth remembering that though our computers fascinate us, consider the organ that's being fascinated itself.

Our brains have about a hundred thousand times the connections in a state-of the-art Cray computer. These connections work about a hundred thousand times faster than the comparable computer networks. This yields an organism with about ten billion times the capabilities of our billion-dollar number-crunchers. Consider what could be done by modifying some of the wiring diagram of that brain, or perhaps just some of its inherent chemistry. The potential for vast improvement or vast damage is immense.

Our currently common idea of software running on hardware works for machines, but not for brains. Brains modify themselves in response to strong inputs, they don't just store data in files. They form new patterns for thinking—self-programming and self-hardwiring. To reflect this, I think we will need a new category—liveware.

Like art, 'living' is a property nobody can define exactly but everybody thinks they can recognize. The virtue of live technologies is the same as the dray horse—it can look after itself, in its own fashion. Cropping grass, relieving itself, burning that grass for energy in its belly, the horse does a lot of its own maintenance. Liveware would similarly police up its own act, and be able to make copies of itself into the bargain, just like the dray.

Of course, a bioteched piece of liveware will be patentable—that's already established in principle—and, alas, mutable. Once made, it can undergo mutation and make something we did not intend.

Intention is the crux of the moral issues we will face. The abortion battles of our day will pale compared with the far more intimate and intricate capabilities that yawn just a decade or two away.

In the USA abortion won't go away as an issue, mostly because we keep trying to settle it through the courts. I suspect the Supreme Court will follow established practice and turn such a hot potato back to the states to decide, as they once tried to do with slavery. But that won't work when changes come thick and fast, as they are starting to.

Already Brahmins in India use amniocentesis to determine the sex of an embryo early on—and then preferentially about the girls, because sons are more prized. This "genetic counseling" frames a typical conflict between our easy categories. Where does 'reproductive choice' end when it systematically acts against females? If allowed to go on, we could produce harrowing population differences far from the near 50/50 balance of sexes, a testosterone steeped society with more crime and war. I don't know the answers here, but I do know that the questions will get tougher.

And more subtle, as well. The first genetic tuning will be for the elimination of inheritable diseases—kidney disorders, hemophilia and the like. Then will come cosmetics: eye and hair color, skin tint, maybe breast size (look at the implant industry today) and height.

These are plenty, but what if parents can tailor their children for beauty? A firm jaw for men, a sunny smile for women? We all know that good-looking people do well. What parent could resist the argument that they were giving the child a powerful leg up (maybe literally) in the competitive world?

Somewhere, law or fashion or deeper arguments will draw the line. But wherever that line occurs, there is the familiar problem of oversubscription. Just as a Bachelor's degree was once a proud emblem, now tarnished by being commonplace, beauty—and maybe even brains—will come to be so. Indeed, since beauty is another form of fashion, generations may sport trendy noses and thighs, as now we see passing fads in children's names.

Of course, the first genetic editing and rewriting will be done for the rich. One of our challenges will be to spread the benefit, or else a class separation will develop of frightful complexity and depth. We could reach the stage in which one could spot the rich by their looks, or even their smarts.

Or their mates. Classical liberalism holds that information is good. And the *truth shall make you free!* Why, then, should a prospective bride not know the precise genetic endowment she would get from a candidate swain? We are just beginning to consider whether a genetic propensity for disease should be made known to insurance companies or employers.

Those legal battles can be settled in the context of property rights. But how about something as intensely personal as marriage? People care deeply about their children. It seems plausible that they would want to know what they are getting before going to the altar.

All these naturally arising problems will tend to make us think of other people as anthologies of genetic traits—to atomize. This reflects science's tendency to slice and dice experience for convenience of analysis, but it is a poor model for knitting up the already raveled threads of a tattered society.

So somewhere, a line must be drawn. It had better be fixed by open public debate, rather than by our current method of leaving it up to lawyers in courtrooms, who usually know little and care less.

Other developments, just over the horizon, will probably force us to entirely rethink present ideas of good and evil. Within a generation, we will probably be able to make cocaine from a bacterial culture. Kids will grow it or morphine or opium or marijuana—in bathtubs, not in elaborate labs.

This will do for our current drug prohibition what home-brewed beer did for Prohibition. Even easier ways are plausible: say, a bacterium which lives in your digestive tract and makes just the right level of cocaine every day. (Something like this has happened naturally. A patient turned up who was permanently drunk from a yeast which made alcohol in his innards.) Far more exotic methods of eluding detection, and of making new designer drugs, will no doubt emerge.

Such a ready supply will almost certainly doom a simple War On Drugs approach. Legalizing, taxing and regulating their use will come to be far cheaper than following a Prohibition mentality against an ever-improving biotechnology.

In fact, I believe it already is cheaper and smarter. We have 1,300,000 in prisons in the USA, the majority for drug-related crime. The average sentence for murder in California is for fewer years (eight) than the average sentence for drug crimes. Batik will make these problems far worse, forcing a new social solution, probably resembling the European solutions already using partial legalization.

Out of playfulness, I've scrambled many ideas together without talking about when they might come. To orient ourselves, I would call 'mundane' the measures which have obvious market roles right away, and little social resistance. This includes pollution-policers, simple bathroom cleaners, crops that resist pests and herbicides, pharm animals, "designer" plants [blue roses, low-cal fruit], bacterial mining, and the like. Even correcting human inheritable diseases will probably go through without major opposition. All this, perhaps within the first two decades of the new century.

The battles will begin in earnest with conceivable but startling capabilities. The list is long. Big changes in our own genome. Harnessing natural behaviors to new tasks (the acacia ant—orange tree marriage). Designer animals, like a green Siamese cat to match your furniture, or a talking collie (and what would it say?). These may preoccupy the middle of the next century.

Even further out would be major alterations in the biosphere, and in us. Adapting ourselves to live in vacuum or beneath the sea, or to convert sunlight directly into energy, would alter the human prospect beyond recognition. Changing *homo sapiens* to something beyond will be a step fraught with emotion and peril. Such issues will loom large as the Biological Century runs out. And what could lurk beyond that horizon? The mind boggles.

All these are mere glimpses of what awaits us. A century is an enormous span, stretching our foresight to the full. Reflect that H. G. Wells' *The Time Machine* was written only a century ago. Batik can usher in as profound a revolution as industrialization did in the early nineteenth century. It will parallel vast other themes—the expansion of artificial intelligence, the opening of the inner solar system to economic use, and much, much more

The Achilles heel of predictions is that we cannot know the limitations of biotechnology until we get there. A nineteenth-century dreamer might easily generalize from the forthcoming radio to envision sending not merely messages by the new 'wireless', but people. Matter, after all, is at bottom a 'message'. But there's more to it than that, and the awesome radio didn't develop into a matter transmitter, which is no closer to reality than it was then.

So undoubtedly I'm wrong about some of these dreams of mine, particularly the timing. What I will bet on is that, despite the current fashion for 'nanotechnology'—artifice on the scale of a nano-meter, the molecular level-batik will come first. It is easier to implement, because the tiny 'programs' built into life forms have been written for us by Nature, and tested in her lab.

In fact, some of the most interesting prospects of nanotech-like thinking come from biological materials. The basic mystery in biology is not how proteins figure but how they fold themselves, which determines myriad biological functions. An obvious long chain molecules to fold and use as a construction material is DNA itself. A self-replicating 'bio-brick' could be as strong as any plastic.

Consider adding bells and whistles at the molecular level, through processes of DNA alteration. Presumably one could then make intricately malleable substances, capable of withstanding a lot of wear and able to grow more of itself when needed.

It isn't fundamentally crazy to think of side-stepping the entire manufacturing process for even bulky, ordinary objects, like houses. We have always grown trees, cut them into pieces, and then put the boards back together to make our homes. Maybe we will someday grow rooms intact, right from the root, customized down to the door-sills and window sizes. Choose your rooms, plant carefully, add water and step back. Cut out the middleman.

Whether such dreams ever happen, it seems clear that using biology's instructions will change the terms of social debate before nanotech gets off the ground.

The rate of change of our own conception of ourselves will probably speed up from its present breakneck pace. The truly revolutionary force in modern times has been science far more than revolutionary politics or the like. That seems likely to be even more true in the future.

Yet the above examples underline the implications of leaving genetic choices to individuals. Perhaps here we see the beginnings of a profound alteration in the essential doctrine of modern liberal democratic ideology. There may be genetic paths we will choose to block.

Our species has made enormous progress through swift cultural evolution. Now that quick uptake on changing conditions can come from genetic change. This tilts the game back to Nature's rules, but with us at the controls, not pitiless, random mutation.

It is as though prodigious, bountiful Nature, for billions of years, has tossed off variations on its themes like a gushing Mozart. Now Nature finds one of her casual creations has come back, grown searching vision, and is eyeing Her oddly.

NINE

PRENATAL GENETIC TESTING AND EUTHANASIA: TWO CASES FOR RESPONSIBLE USE OF FREEDOM

Stephen Post

Even in a pluralistic culture it is possible to accept common minimal principles of ethics such as nonmaleficence, autonomy and justice, but when these universal principles are applied through the filters of many interpretive worldviews, considerable moral relativism persists. For example, we may all agree on the principle of nonmaleficence ("do no harm"), but disagree on whether this principle applies to the human fetus at birth, at quickening, at viability, or even at birth. We may disagree about who is dead and therefore no longer under the protective umbrella of nonmaleficence, for some accept whole brain definitions of death, and some cling to traditional views (e.g., Orthodox Jews and most Japanese). We may hold to egalitarianism with respect to sentient species, so that our eating the meat of nonhuman animals is judged pernicious. My point is obvious: consensus about a principle of ethics does not eliminate moral relativism, due to the central place of interpretation in valuing life. The most we can sometimes hope for is a civil conversation across traditions of disparate interpretation, one that avoids acrimony and distortion.

In a free society, that is, one that understands ground gained for individual conscience as ground appropriately lost to the state, there are few moral positions that should be widely enforced. Law exists to establish the minimal restrictions on behavior that prevent harm and separate social experience for the "war of all against all." It neither enforces moral idealism nor resolves many of the quandaries of human conscience. Yet it is still advisable that we discuss as free individuals the moral and ethical arguments available so that our freedom is informed and competently expressed. The libertarian should find no reason to lament such considerations so long as choice is left to personal conscience.

In the field of biomedical ethics, there are at least two major issues for the 1990s: genetics or the Human Genome Project, and euthanasia or mercy killing. With respect to genetic information, it has been argued that anyone should have any and all information he or she desires in the context of both prenatal and presymptomatic carrier screening. With respect to euthanasia, it has been argued that any competent human being has a right to decide on the moment of death, and to implement said desire either through assisted suicide or mercy killing. I will suggest that in both these areas ideological distortions occur.

THE HUMAN GENOME PROJECT AND GENETIC TESTING

The project to map and sequence the human genome is driven by the hope of gene therapies to ameliorate or cure human diseases. Almost weekly the media reports a new discovery such as the gene responsible for congestive heart failure, Huntington disease, or familial Alzheimer disease. These discoveries are heralded as potentially important therapeutically. The National Center for Human Genome Research of the National Institutes of Health (United States) is engaged in a worldwide research program known as the Human Genome Project. Its goal is to map the entire DNA content of the human being.

One concern is with the distortion known as genetic determinism, which overstates the extent to which human disease is genetically determined. Since the late 1950s, a rapid expansion of knowledge as to how genes work has resulted in the myth that through genetics utopia is on the horizon. For example, despite weak evidence, the "criminal chromosome" myth captured the public imagination. But it was soon clear that the XXY genetic configuration resulted neither in hyperaggression nor criminal behavior. However, the XXY myth was already embedded in high school and medical school texts.

Increasingly, the focus on the human genome in basic science and in the media is shifting international attention to both genetic explanations and solutions of medical or social problems. Some scientists have suggested that the Human Genome Project will solve the problem of homelessness. The presumptions are that homelessness is due to mental illness, that all severe mental illness has a genetics basis, and that by discovery these genes cures can be developed. The field of psychiatry, long on the defense in medicine because of its empirical softness, is increasingly focusing research on genetics. This does not mean that mental illness is no longer considered multifactorial, or that psychotherapy is outmoded; yet the quest is for a clear genetic indicator and a genetic solution. Increasingly, we see the medicalization and geneticizing of social problems rooted in family disfunction or poverty.

What we await is any clear evidence that gene therapy will work. Moreover, we seldom hear of the morally questionable aspects of gene mapping. How can therapy be clearly distinguished from enhancement, and how likely is it that medical science will avoid marketing new images of human perfection? The therapeutic repair of human beings is noble, but efforts to enhance the already healthy are inherently problematic. What defines enhancement? Are taller or more slender people better? And where would the endless so-called enhancement end? Serious and objective medical need, rather than the vicissitudes of enhancement, are the proper basis for genetic interventions. The possibility for confusion between mere human *wants*, and genuine human *needs*, is always real. A parent may want a "designer" child via gene enhancement, but this is not something that parent or child needs.

Additionally, mapping will dramatically increase the number of gene abnormalities that can be tested for to many more within the decade, and perhaps eventually to thousands. While selective abortion is an old topic, the genome project raises it in a newly important way. Many of us accept prenatal diagnosis and subsequent abortion for grave or relatively serious genetic defects that will manifest early in the sufferer's life, but are critical of termination of pregnancy for trivial or moderately serious genetic indications, and for indications that will manifest only later in life. Even on the premise that there is a basic right to elective abortion, of which the right to selective abortion is a subset, there is, nevertheless, room to discuss the moral underpinnings of the choices that women are free to make. Barbara Katz Rothman provides a number of case studies in which abortion was chosen as an alternative to a child with mild diseases and less than disabling impairments. She suggests a future in which we will see "a rise in the standards of production for children" that emerges from new technologies. "Will we," she asks, "establish a set of norms of acceptability, and then narrow, and narrow, and narrow yet again those norms" (1986, p.227)?

In response to this potential problem, Dorothy C. Wetz

and John C. Fletcher, supporters of the genome initiative, argue that the medical profession should abandon a position of ethical neutrality with regard to prenatal sex selection, partly because this sets precedents for selective abortions unrelated to disease or disability, for example, eye and hair color, thinness, skin color, straight teeth, and other "cosmetic" considerations. Within a decade or two, they continue, these "exotic" choices will be technically possible, especially relating to body size and height (1989).

Prenatal testing will eventually be capable of detecting hundreds or thousands of single gene defects, many more polygenic and multifactorial defects, and numerous superficial characteristics of aesthetic concern. Pregnant women, at about eight to ten weeks gestation, may be able to have a blood test indicating the DNA profile of the fetus bases on fetal cells in the maternal circulation. This extensive new level of knowledge leads to tremendously complex personal choices about what lives are worth living, qualitatively considered. That women have the right to choose abortion is widely accepted in American culture, but the discussion only begins here respecting personal moral conscience in the throe of decision.

In the absence of an obviously grave and immediately threatening defect, vexing decisions will be made based on severity, probability, and age of onset of disease or disability. Adult onset polycystic kidney disease, which may or may not occur, and which results in progressive renal failure during the adult years, is treatable by dialysis or transplant. In this case, moderate severity combines with uncertainty of manifestation and late onset. Huntington's disease can be distinguished from adult onset polycystic kidney disease because it is much more severe, and untreatable. Would an abortion be morally justifiable for a fetus if the future child has a 20 percent probably of bipolar affective disorder or schizophrenia? What about familial Alzheimer's disease? What shall we do with the freedom to decide, especially when genetic conditions have variable expression from mild to serious, variable likelihood of manifestation, and variable age of onset?

At a minimum, we can distinguish moral from aesthetic values, and give priority to the former. A disease such as Huntington's may be insufficient grounds for selective abortion because, even though it is clearly very severe, the eventual sufferer nevertheless will have many decades of good and unimpaired living. Moreover, the parents of the child are not immediately or even directly affected in the way they would be were the disease of early onset.

I do not want to go very far in resolving the balance between severity, probability, and age of onset that might justify selective abortion. Rather, I offer several humanistic reflections to provide a general background for such decisions, and that together justify reservations about abortions for disease of late onset, such as Huntington's or possibly familial Alzheimer's. My limited intention is to comment on American culture, focusing on three themes: the parental desire to avoid bringing suffering into the world; the contingencies of the human condition; and the moral ambiguity of the quest for "perfect" babies. These themes will be linked to both selective abortion and, more briefly, to gene therapy.

Suffering

Parents rightfully prefer not to bring lives filled with suffering into the world. Few, if any, would quarrel with the assumption that it is preferable to have healthy children who are not born into physical pain. When prenatal diagnosis reveals a grave defect that makes life an onerous burden of suffering, nonmaleficence warrants abortion. But it is wrong to assume that suffering is the necessary result of genetic defect, or that lives with degrees of physical suffering cannot be creative and meaningful.

One advocate of rights for disabled persons points out

that as prenatal diagnosis results in vast new genetic knowledge, women need "to obtain far more and different information than they very commonly get about people with disabilities" (Asch, 1989).

The notion that all disabilities cause suffering is conceptually flawed. In many cases, negative stereotypes obscure the creative ways in which people with disabilities cope with different challenges and needs.

The Human Genome Project calls for scrutiny of the assumption that those who are *different* necessarily suffer. With our societal inclination to rather rigid standards of beauty and physical prowess, self-reliance and productivity, it is too easily assumed that those who fall short of these standards therefore suffer. Compassionate discrimination, which makes the experience of genetic impairments out to be worse than it is, should be avoided.

The desire to eliminate disease and the sufferings that may be associated with illness is morally valid. However, the definition of suffering is wrongly expanded to include the ways in which an individual is different from others, though fully healthy. Suffering becomes a social construct imposed on us, so that parents will petition the physician to "enhance" a child regardless of the onerous imposition on the "patient" and the folly of the request. It is incumbent on physicians to hold firmly against the quest for enhancement, in part by maintaining a disease-based definition of the human suffering for which medical therapy is responsible. To widen the definition of suffering so as to provide enhancement interventions is precisely the wrong response to the human condition. Moreover, such interventions violate the purpose of the healing art, which is the restoration of physical and mental function when possible.

Contingency

Human experience is partly uncontrollable, and therefore contingent or chance-ridden. Our desire not to bring suffering into the world should be tempered by a recognition that suffering is a part of life, and escapes human prevention to a large degree. Those who are genotypically and phenotypically more "perfect" than others can lead tragic lives, however much we try to prevent this. Take the case of the great French artist Henri de Toulouse-Lautrec. A descendent of aristocrats, he was the victim of two accidents which broke his legs and left him incurably disabled. His torso developed but not his legs, and he became deformed, unable to walk without a cane. He derived some consolation from painting, until dipsomania led to the asylum.

His was an irregular life, one of immense suffering; it was also one of creative compensation and the development of the artistic poster as we know it today. Toulouse-Lautrec was born a normal infant, for all intents and purposes, a perfect baby. But the contingencies of human experience that range from accident to bad luck left him disabled anyway, and suffering from a diminutive stature. The diminution sadly befell him but simultaneously may have elevated him artistically, although it is erroneous to suggest that disabilities generally give rise to unusual forms of creativity as a compensatory response.

The classical pre-scientific western culture left events largely in the hands of a mysterious deity whose ultimate purposes were presumed loving. Does technology foster a rage to control, and prevent our coming to grips with the basic existential reality of contingency from which we never fully escape?

No matter how much we attempt control, suffering is a part of all human lives, to greater or lesser degrees. The Tantric Buddhist will state that if you are born, you will suffer; life is understood as suffering unto growth, until the wheel or rebirth is finally escaped. Indeed, transformative suffering is viewed as the only point of human bodily existence.

Perfection

The right of only so-called perfect babies to exist is not a matter of public policy, but each time a selective abortion for a moderate or trivial imperfection occurs, we are in effect accepting this principle. Gradually, society moves perilously closer to Nietzsche's norm from *The Anti-Christ:* "The weak and ill-constituted shall perish: first principle of *our* philanthropy. And one shall help them to do so."

All perfectionism must be tempered by an awareness of what Leslie A. Fielder dubs "the tyranny of the normal." Fielder notes a "deep ambivalence toward fellow creatures who are perceived at any given moment as disturbingly deviant, outside currently acceptable physiological norms." He refers to "a vestigial primitive fear of the abnormal, exacerbated by guilt." Fielder fears the "enforced physiological normalcy" that sent dwarfs to extermination camps in Hitler's Germany. "Perhaps it is especially important for us to realize that finally there are no normals, at a moment when we are striving desperately to eliminate freaks, to normalize the world" (1985).

One of the ways in which persons who depart from "normals" contribute to the community is by challenging us to overcome social stigmas, and to accept difference in our midst. Views of physiological human perfection are inevitably intertwined with stigmas, one form of which is abominations of the body. Those whose bodies depart negatively from the "normal" are the victims of a socially shaped tendency to revulsion. Stigmas specific to the body are as morally problematic as those related to religion, race, and nationality, and often cause great suffering to disabled people. People who are different and "imperfect" teach us about the meaning of equality and commitment. But we are beings who fear difference, so diversity is hard to sustain.

The very nature of human perfection has, of course, been the subject of acrimonious debate over the centuries. In the medieval period, there was a profound sense that perfection is chiefly a matter of character and virtue, and that bodily imperfections provide opportunities for concentration on the internal moral and spiritual values. Indeed, the weight of religious symbolism, from the club-footed Christ figure of the Eastern Orthodox icons to Dostoyevsky's idiot epileptic savior, underscores the inward perfection made possible by external limitations.

There is treasure in earthen vessels, and earthen vessels we humans are, subject to countless infections, accidents, chronic ailments, and finally to the decline of old age and death that we in this culture try so hard to deny, as though senility were mere myth. Arguably, our culture focuses perfection on the vessel rather than on the person within it. Of course, it is reasonable to avoid bringing grave human imperfection into this world. Infants with no relational potential should not be born. But we must be highly circumspect about declaring too imperfect those who must endure somewhat earlier in life the very sorts of frailties that eventually assault each one of us.

It is especially ill-conceived when a society so overvalues beauty and physical prowess that ugliness and bodily weakness are aborted out of existence. Aesthetic vicissitudes might increasingly determine who should, and who should not, inhabit the world. But this determination is fundamentally flawed (the Buddha would laugh). It is rooted in mistaken attachments to the bodily container of the human self, and not to the inner self. Arguably, our culture is pitifully narrow in the externality of its perfectionism. We must reflect on the abyss of racial hygienics.

In particular, enhancement and eugenic genetic engineering are problematic because they further externalize our images of human perfection, and do not result in any clear moral good. By externalize perfection I mean definitions of the human good that are centered on the shape of the body, or on some particular capacity for music, visual arts, and so forth. All the major cultures of the world have defined human perfection internally, that is, with emphasis on character and virtue. From Aristotle to Thomas Aquinas, perfection meant wisdom rooted in experience and in the relationships by which the moral life is learned through example. Our perfection lies not in gene enhancement, but in the enhancement of character.

THE ISSUE OF EUTHANASIA

A good death is no longer typically defined in terms of retrospective self-assessment and repentance, or of customary religious rituals calling together family, friends, and neighbors. A good death, "euthanasia," is increasingly defined as a direct, active, and voluntary preemptive strike against decline and dependence on others. The active, directly intended, and freely chosen self-destruction that was once unspeakable has become speakable. The debate in the United States over this issue has been in large part focused on the "slippery slope" arguments, for example, that voluntary euthanasia will lead to nonvoluntary euthanasia (the killing of the severely demented patient, the patient in the persistent vegetative state, and so forth). This, in turn, will lead to involuntary euthanasia, that is, killing people against their will. At the center of this ideological flurry is the Nazi analogy. Any physician who assists in a voluntary patient suicide is considered a Jose Mengle waiting to happen. My view is that the Nazi analogy is used to grease a slippery slope that is, in fact, not terribly slippery at all. In other words, it is reasonable to think that certain clear steps could be built to prevent the fabled slide into the moral abyss. This does not mean that I advocate assisted suicide or mercy killing, but that I do not believe that slippery slope arguments and Nazi analogies are very persuasive.

The obstacle to the progress of the euthanasia movement in Great Britain and America was the abyss of the Nazi eugenics movement, and the appalling revelations of medical killing at the Nuremberg Tribunal. Yet now, with Holland the

Royal Dutch Medical Association placing *de facto* imprimatur on euthanasia, with the growth or the Hemlock Society (and its political wing, Americans Against Human Suffering), with revelations in major medical journals by physicians who have killed their patients, with Dr. Jack Kevorkian's suicide machine, and a 1990 *Time/CNN* poll indicating that 57 percent of Americans approve of physicians administering lethal injection to unconscious terminally ill patients who have indicated this preference by advance directive, the times are changing.

My response to this movement toward euthanasia is cultural, rather than philosophical-analytical. I will comment on the themes of control, loss of care, and the desire to put an end to human suffering. These themes are not unrelated.

Control

Modern technological culture encourages ever greater control over human events. American families often want their loved ones to die in the controlled environment of the medical intensive care unit. The beeping signals and flashing lights of the machine signify a mastery over nature and human nature. For many, these machines define what is the best standard of care, and any shift away from this is downgrading and therefore undesirable. So frequently, families will resent the offer to move a dying patient from intensive care to a special care unit that provides care and comfort only. Americans often think that if they do not avail themselves of the latest technology, they are certainly "missing out."

The idea of throwing in the towel, of only leaving death in the hands of nature or of a wisdom that underlies nature, is anathema to the rage for control. We witness the same rage in our era of the "perfect" baby, and with the advent of the human genome initiative, selective abortion of all but one's idiosyncratic aesthetic image of a cosmetically ideal child will be increasingly possible. Some will want to control everything from hair color to height. DNA profiles of the fetus will be available based on fetal cells in the maternal circulation. Researchers want to control the aging process with growth hormone or scavenger cells. From the womb to the tomb, technological control is the cultural mandate. With the *ars moriendi,* it was the dying person's *internal* control over a rite of passage that brought order. This inward control has largely given way to mechanical control.

Death by lethal injection is best understood as a further act of technological control. It is driven by the will to control, to remove human events from the domain of nature, even when suffering can be mitigated in almost all cases by proper palliation. Against this form of control there is only the sense that the hour of death is rightly decided by a wisdom beyond us. To return death to the state of nature requires the assumption that underlying nature is the wisdom of God, of some higher purpose or regulation. Your ordinary Americans who oppose euthanasia appeal not to philosophical arguments, but to straightforwardly theological ones.

The technological control we do need is in pain control. New modalities of palliation that leave the patient in a reasonably clear state of consciousness are merging. Electrodes implanted into brain or nerve can modulate pain pathways. Through surgical and cryonic neuro-ablation pain can be managed in remarkable ways. Our approach to pain management must be physical, psychological, multidisciplinary, and spiritual. But in our curative and rescue-oriented health care system, management of pain is *not* a priority. Ignorance of proper pharmacologic principles is pervasive, and physicians are often ill-equipped. Medical schools do not include pain management in their core curriculum. Narcotics are administered on a pain-contingent basis, rather on time contingency so that we wait for the patient to express pain. This is ethically wrong.

Caring

To a degree, euthanasia is also a result of a culture that

devalues caring. Our medical system concentrates on research and training that steals people back from death. It is rescue oriented in the extreme. Merely caring suggests "acting by default." In a highly compassionate call for the recovery of caring, Daniel Callahan writes, "At the center of caring should be a commitment never to avert its eyes from, or wash its hands of, someone who is in pain or suffering, who is disabled or incompetent, who is retarded or demented; that is the most fundamental demand made upon us" (1990, p.145). Callahan is appropriately critical of our failure to train medical students to care in the seminal sense of the word, and he is also right in his pessimism that however much care is discussed in glowing terms, "it always loses out to an emphasis on scientific knowledge and technical skills, and there is no end in sight to that bias" (p.147). Technology has "muscled aside" the most basic expressions of care.

It was not always so. Once, there was little the nurse or physician could do but hold the dying patient's hand, usually at home, and with some religious motivations. Now, we need to recover such simplicity once rebellion against dying becomes futile.

The caring that the dying really need, other than palliative, is a compassionate response, an enduring supportive emotional intimacy. In a culture where passion is more highly valued than compassion, the tasks of caring are readily viewed as demeaning.

Partial or constant dependence on others is viewed as an unreasonable imposition, and as personally demeaning, never mind that caring is a basic human need. Euthanasia has an appeal when the forms of caring that make it unnecessary are systematically de-emphasized and devalued.

Suffering

A culture of external, as contrasted with internal, control merges with the loss of caring in the most basic sense of the word. A third ingredient is this: American culture is extreme in its desire to eliminate suffering. It is remarkable, if Aries is right, that in earlier time suffering was accepted as much as appears to have been the case, despite the absence of palliative methods. Now, utilitarians speak of the "greatest Happiness of the greatest number," and our advertisements inculcate the notion that to be zestfully happy is to be acceptable and good. Morbidity and the decline of dying are antithetical to the dominant image of human normalcy, for they require us to face our limits. Euthanasia is conceptually flawed if the motive is the elimination of suffering. Human beings always have and always will suffer. The Tantric Buddhist maintains that if you are born, you will suffer; life is understood as suffering unto growth, until the wheel or rebirth is finally escaped. Christianity makes the cross its central symbolic. In a culture that can no longer depend on religious recognitions of human finitude, decline and suffering must be removed from the scene.

Will euthanasia win over the western world? Or will we look for better ways to control pain, explore a wider range of forms of supportive care, and allocate resources so that long-term care does not require a spend down into poverty? Acts of assisted or unassisted preemptive suicide and euthanasia are one possibility. The other possibility is to understand mercy not in terms of killing, but in terms of providing care. For all the vast literature on euthanasia, one of the most persuasive pieces remains Arthur Dyck's "Beneficent Euthanasia and Benemortasia: Alternative View of Mercy." Dyck stresses pain relief by dosages of medicine sufficient to accomplish this purpose; relief of suffering through companionship; a patient's right to refuse treatment; and a health care system that does not impose such financial pressure on patients or their families that euthanasia becomes attractive. He also emphasized that if our culture so desires, "mercy" can mean loyal compassionate long-term caring of the dying, rather than merciful killing (1975).

But what we do not need is the promiscuous appeal to

Nazi analogies, or suggestions that those who conscientiously adhere to assisted suicide or mercy killing are pernicious. If Aristotle was right, ethics is an inexact science permitting considerable moral disagreement. While I remain skeptical of a policy of euthanasia (but I accept certain cases where the act would be moral, e.g., in the absence of palliation), I find it impossible to condemn those who would extend autonomy to include euthanasia. The ancient Stoics thought such self-deliverance fitting.

REFERENCES

A. Asch "Can Aborting "Imperfect" Children be Immoral?" in A. Arras and N. Rhoden, edsl, *Ethical Issues in Modern Medicine* (Mountain View, California: Mayfield Publishing, 1989), 319-21.

L.A. Friedler, "The Tyranny of the Normal," in T. H. Murray and A. L. Caplan, eds., *Which Babies Shall Live? Humanistic Dimensions of the Care of Imperiled Newborns* (Clifton, New Jersey: Humana Press, 1985), 151-59.

D. Callahan, *What Kind of Life: The Limits of Medical Progress* (New York: Simon & Schuster, 1990).

A. Dyck, "Beneficent Euthanasia and Benemortasia: Alternative Views of Mercy," in M. Kohl, ed., *Beneficent Euthanasia* (Buffalo: Prometheus Press, 1975), 117-29

B. K. Rothman, *The Tentative Pregnancy: Prenatal Diagnosis and the Future of Motherhood* (New York: Penguin Books, 1986).

D. C. Wertz and J. C. Fletcher, "Fatal Knowledge? Prenatal Diagnosis and Sex Selection," *Hastings Center Report* 19 (1989); 21-27.

TEN

O BRAVE NEW (VIRTUAL) WORLD

Ben Bova

It's the hot new topic in high tech. There have been articles about it in magazines ranging from *Time* and *Newsweek* to *Smithsonian* and *Omni*. Television shows have been done about it. Now there's a new book devoted to the subject of virtual reality.

For me, reading Howard Rheingold's *Virtual Reality* was like watching one of my children leave home and go off into the world. I have been writing about VR for thirty years. Of course, my writings on the subject have been fiction. Now the facts are catching up with me.

In 1962 I wrote a short story titled, "The Next Logical Step." It was published in *Analog Science Fiction* magazine for the excellent reason that it was a science fiction tale, a story about new technology and its effect on the human race. No "straight" magazine would touch it; after all, it was thirty years ahead of the real world. It dealt with an arcane machine developed by the Department of Defense to help military planners visualize the wars they were preparing to fight.

"This is the most modern, most complex and delicate computer in the world," one of the characters tell another: "It was built to simulate actual war situations. We fight wars in this computer... wars with missiles and bombs and gas. Real wars, complete down to the tiniest detail. The computer tells us what will happen to every missile, every city, every man...who dies, how many planes are lost, how many trucks will fail to start on a cold morning, whether a battle is won or lost...."

So far, the story was hardly science fiction. The Pentagon had war-gaming computers in the early 1960s, though none of their programs were as detailed as my fictional machine's.

"Yes, but this machine is different," my narrator continues. "We've added a variation of the electroencephalograph... a recording device that reads the electrical patterns of your brain. Like an electrocardiograph. But you see, we've given the EEG a reverse twist. Instead of using a machine that makes a recording of the brain's electrical wave output, we've developed a device that will take the computer's readout tapes and turn them into electrical patterns that are put *into* your brain!"

I had invented a VR machine.

One of the generals in my story explains, "You sit at the machine's control console. A helmet is placed over your head. You set the machine in operation. You *see* the results.... You actually see the war being fought. Complete visual and auditory hallucinations."

The concept that would eventually be called virtual reality continued to haunt me. In 1969 I developed a variation in the novel *The Dueling Machine.* This time the VR device allowed two people to share an imaginary universe. "They can do anything they want in their dream world," says the devices inventor: "Settle an argument as violently as they wish, and neither of them is hurt any more than a normal dream can hurt you physically. Men can use the dueling machine as an outlet for their aggressive feelings, for their tensions and hatreds, without hurting themselves or their society."

Can they? In my novel, evil men begin using the dueling machine to commit real murders. In the world we live in today, will actual VR machines be used to make our lives better, to further knowledge and understanding, or merely to make money no matter what the consequences to society?

Virtual reality may become a powerful influence in our lives. Through VR technology we will be able to convince our bodily senses, and perhaps even our minds, that we are seeing, touching, hearing, tasting, smelling environments that actually exist only in a computer's chips. We will be able to leave the real world behind, temporarily, to experience worlds of imagination or illusion with all our senses. The distinction between reality and imagination may become blurred.

Rheingold examines these challenges and opportunities at some length. But first he gives a thorough description of what virtual reality is, how it works, and who its pioneers and missionaries are.

Three separate strands of technological development converged over the course of a generation to bring VR out of the pages of science fiction and into reality. One of those technologies is what Rheingold calls "the experience theater": devices that show you pictures or sounds or other sensory inputs. The inventors of Sensorama and Cinerama were among the grandfathers of VR. The children who haunted video-game parlors were among VR's earliest experimental subjects.

Eventually the engineers developed stereoscopic goggles that are actually miniature television screens capable of showing you pictures in three dimensions. And "data gloves": instrumented gloves that can sense the motion of your individual fingers or your whole hand and transmit that information to a computer, where the data will be used to manipulate tools or other objects, or to change the pictures you see. The remote manipulators that nuclear technicians call "waldoes" were the predecessors of today's VT data gloves.

Then there was a group of offbeat scientists and engineers whom Rheingold calls the "infonauts." Some of them worked for government or university laboratories, and others worked in private industry, but they all seem to have had one trait in common: They were dissatisfied with the way computers were being developed. A decade before personal computers entered our lives and the term *user friendly* became a catchword in the business, these men felt that computers would never reach their full potential as long as their users had to learn arcane computer languages to operate the machines.

In those distant days (less than twenty years ago!), dedicated computer technocrats were sometimes called "chip monks." They believed with nearly religious fervor that computers would solve all the world's problems, and that learning the secret languages of computer programming was a sort of initiation rite into their tightly bound world of hope and promise.

The infonauts believed just the opposite. Each of them wanted to create simpler, more human ways for people to use computers. Instead of expecting users to learn the computer's operating language, they wanted to develop systems that would allow an ordinary man or woman, or even a child, to operate a computer using everyday language or picture-like symbols that are immediately understandable—or even hand motions, like the conductor of an orchestra.

While the infonauts were striving to make computers easier to use, another group of technologists was wrestling with a different problem: simulation.

Before World War II the aircraft industry began to realize that it was cheaper—and far safer—to train new pilots in machines that *simulated* flight rather than allowing them to go up in a real plane and take the chance of killing themselves. The World War II era of simulators were crude devices. But they did save lives and the cost of crashed airplanes.

As aircraft became faster, more complex, and more expensive, simulators became more sophisticated. Space flight made accurate simulators even more necessary. When astronauts were asked to "go where no man has gone before," they needed to know what to expect of their spacecraft and the alien environment in which they would be operating.

Meanwhile, war itself was becoming so complex that military planners needed detailed scenarios of "what would happen if." My early short story was based on the computer war games that were being developed in the 1960s to help strategists foresee the consequences of their plans and actions.

Behind these three lines of technology was the continuing improvement in the electronic digital computer itself. The multi-million-dollar mainframes of the 1950s, which took up whole rooms and needed teams of specialists to care for them, became the microcomputers of the 1970s and the desktop minis of the 1980s. Computer power increased and price decreased. The standard wisdom among computer people is that if automobiles improved their performance and dropped their price as much as computers have over the past thirty years, you would be able to buy a Rolls Royce for a hundred dollars and it would give you a hundred miles to the gallon.

Powerful, fast computers. Simulations that can show you what it's like to walk on the moon or fly a hot new jet. Operating systems that respond to the movements of your hands and eyes.

Virtual reality. Just who coined the phrase is uncertain. It has also been called "alternative reality." Science-fiction author William Gibson dubbed it "cyberspace in his 1984 novel *Neuromancer.* Virtual reality has become the catchword, the title that is most generally used to describe the machine-produced environment in which you can experience a complete electronically induced hallucination.

"At the heart of VR is an *experience,*" Rheingold says, "the experience of being in a virtual world or a remote location...."

You are in a "virtual" world of the ultrasmall. You see individual atoms floating all around you. You grasp two and try to link them. They resist your efforts with a force based on their electrostatic potential—those valences that you struggled through in high-school chemistry. Through trial and error, in minutes, you can determine which chemical compounds can be produced and which will not work.

Imagine yourself a surgeon performing a delicate operation to remove a cancerous brain tumor. Fiber-optic probes with miniature television cameras look inside the patient's skull. With a VR system you are linked to those probes; as far as your senses can tell you, you are inside that skull, clearly seeing healthy brain tissue and the diseased mass of the tumor.

You are also linked to the surgical laser that will burn away the cancerous tissue when you command it to do so. You touch the cancerous mass with your extended finger, destroying the tumor cells with the godlike touch of your hand.

Picture yourself as a researcher seeking to develop a new plastic that will be biodegradable. Instead of working in a laboratory glistening with glassware and bubbling chemicals you stand in a seemingly empty room, your head encased in stereoscopic goggles, VR gloves on your hands.

Walk on the untouched sands of Mars. Explore the bottom of the ocean. Dismantle the radioactive core of a nuclear reactor. VR allows you to go wherever our tools and machines can travel. Indeed, wherever our imaginations can travel. In your mind's eye, in your own brain's sensory receptors, you are there—even though your body may be sitting calmly in a room, encased in VR goggles and gloves.

Then there is the possibility of what Rheingold calls "teledildonics": VR sex. Instead of merely a glove that responds to the movements of your hand, how about a complete bodysuit that not only responds but produces sensory stimuli? A lover's soft caress, even though your lover may be a thousand miles away. Or without needing an actual lover at all. VR can deliver the man or woman of your dreams, the current sex object of the cinema, famous persons of today or yesteryear. The prospects are virtually endless and endlessly fascinating.

Disturbingly so. Rheingold even claims that "when today's infant VR technology matures in a few years, it promises (and threatens) to change what it means to be human."

Will VR truly have that enormous an impact upon us? Will virtual reality become "Electronic LSD," as the *Wall Street Journal* has called it? Will the children of yesterday's video-game addicts become addicted to their own private electronic realities, lost in a sort of Dungeons and Dragons maze created by their own imaginations and the delicate intricacies of VR systems? When you can pull on a VR helmet and gloves, or a whole bodysuit, how powerful will be the temptation to "tune in, turn on, and drop out"? Timothy Leary is a big fan of virtual reality, apparently.

Rheingold is quite aware of these pitfalls: "One way to see VR is as a magical window onto other worlds, from molecules to minds. Another way to see VR is to recognize that in the closing decades of the twentieth century, reality is disappearing behind a screen. Is the mass marketing of artificial reality experiences going to result in the kind of world we would want our grandchildren to live in?"

The Genie is out of the bottle, and there is no way to reverse the momentum of VR research; But these are young jinn, and still partially trainable. We can't stop VR, even if that is what we discover is the best thing to do. But we might be able to guide it, if we start thinking about it now.

On the other hand, think of how powerful a teaching tool VR might be. Imagine learning chemistry by walking among the atoms and molecules and *feeling* how they resist your attempts to bring them together. Imagine learning history or geology or engineering by "being there."

It is possible to assess the history of the human race in terms of our ability to acquire, handle, and share information. In paleolithic times, human hunters succeeded not because they were larger than lions or stronger than bears. They succeeded because individual humans could learn how to make tools and then share that information with others.

Anthropologists can trace the spread of toolmaking across whole continents. Historians trace the spread of new ideas around the world. Our bodies have not changed much in the past hundred thousand years, but how our tools have changed! How much information we have acquired! *That* is what has moved us from cave-dwelling nomadic hunters to a globe-spanning civilization in the blink of a geological eye.

Virtual reality is a new way to handle and share information. Will it lead to greater knowledge, greater wisdom? The potential is there. But like all new discoveries, the potential for great harm is present also.

Recalling the "original" VR device in my 1962 short story, the machine created visions so real and compelling within the minds of the generals who used it that they could no longer command armies. The images of death and destruction were so overwhelming that they were literally unable to send troops into battle. Would that be a good thing for society, or an evil?

Written in the depths of the Kennedy-Khrushchev confrontations of the early 1960s, "The Next Logical Step" of the story's title was for the CIA to make certain that Soviet spies got a copy of the machine so that *their* generals would become incapable of fighting a war, too. Will actual VR technology make the world more peaceful? Will politicians and diplomats and even generals be able to share their dream, their visions, their deepest fears? Or will VR technology become still another way for political leaders to manipulate their people?

My inventor of the dueling machine wanted to allow people a chance to work out the frustrations and tensions of modern life by fighting imaginary duels in a VR device.

Would such therapy help people today, or would it merely add to our heritage of violence?

Rheingold has a positive view of the future of VR: "As with other technologies, cyberspace is not an either-or case. It will be both-and. People will use it as a hybrid of entertainment, escape, and addiction. And other people will use it to navigate through the dangerous complexities of the twenty-first century. Let us hope it will be a new laboratory of the spirit—and let's see what we can do to steer it that way."

Rheingold's prose is not always as exciting as the subject matter, but then the subject matter is truly the stuff of dreams coming true.

PART THREE

INITIAL SCENARIOS

ELEVEN

REPORT ON PLANET EARTH: A PERSPECTIVE FOR 2044 A.D.

Charles Sheffield

In the past, they said it could only get worse.

The professional futurists, peering into their cloudy crystal balls back in the 1990s, could see many ways for the world to go downhill. Worse, yet, the decline they perceived would continue as far into the future as they dared to look.

The bogeymen of our grandparents came in two main categories. First, there were impending shortages. The world, they said, would run out of many basic materials in the next half century; cheap oil would disappear by about 2030, natural gas by 2050, low-sulfur coal would be gone by 2090. On the minerals side, although there was enough iron for centuries, lead would show a severe shortage by 2040, and copper by 2050.

Even more alarming, perhaps because less obviously a resource until recent times, the world's demand for fresh water would far exceed its supply by 2020 or earlier. The world's tropical forests would be gone by that date, and the great boreal forests of the Northern Hemisphere would be in decline because of acid rain and the "forest dying" that by the mid-1980s was already affecting half the woods of Europe.

Along with forest clearing and the overexploitation of agricultural lands would come the loss of topsoil, increased desertification, and the consequent decline in the area of arable lands. Food production potential would diminish.

Removal of forest cover, especially tropical forests, would lead to another and irretrievable loss of *species.* In 1990, about three-quarters of the roughly five million species of plants and animals on Earth could be found only in the tropics. In the decade from 1990 to 2000, about a million species became extinct because of deforestation. Another two million would go by 2020.

The projected shortages were alarming. But perhaps more frightening was the second class of problem: the projected *surpluses.* Air pollution by oxides of nitrogen and sulfur would be on the increase. The water supplies would be increasingly contaminated by harmful toxins. The loss of the world's forests would mean an increase in the amount of atmospheric carbon dioxide. This amount would be further augmented by the burning of fossil fuels. The globe would retain more solar heat and induce an overall "global warming" of anything from 1-5°C. The polar ice caps would melt. Sea levels would rise and inundate the world's coastal plains. Arable land and cities would vanish. The hard-core pessimists even thought it possible that Earth might move away from its eons-old heat balance and transformed into a hot and lifeless hell like Venus.

All of these ominous shortages and surpluses were driven by one overriding "Surplus": the human population. From a mere billion in 1800, it had increased to two billion by 1930, to four billion by 1975, and to six and a half billion by the year 2000. Projections for 2050 were frightening: they ranged from a low of eight and a half to a high of 15 billion of more.

Our grandparents examined the trends, and made their gloomy projections. A world of desperate shortages, diminishing options, and degraded life-style seemed inevitable for all but a fortunate few. By 2042, or much earlier, starvation and deprivation would be the norm. What they could not see, although with the benefit of hindsight we can discern clearly, were the seeds, vigorously sprouting, that would transform the world before the middle of the next century.

Since the beginning of civilization, it is likely that people in every era have considered the 50 years immediately preceding their own time as being of unique importance to the world's history. This tendency toward "temporal chauvinism" should perhaps be deplored; and yet we cannot help asserting that the period between the year 2000 and today has truly been the most critical ever for humanity. For it is now true that for perhaps the first time in the whole of human history, the future looks bright.

The most important event of the past 50 years has arguably been not change, but *constancy*. And for the first constancy, we humans can take no scrap of credit. The stability of the vast self-regulating entity that forms Earth's biosphere has proved to be truly extraordinary. Any increase in carbon dioxide levels is followed, almost at once, by an even stronger increase in plant activity. Plants gobbled up carbon dioxide as fast as it could be produced by the burning of fossil fuels. They increased their own growth rates even faster. During the last decade of the twentieth century and the first one of the present, plant growth everywhere showed increased vigor. There was increase in the biomass available for food, fiber, and fuel. Just as important, stimulation of plant root activity accelerated the availability of humus and the renewal of topsoil.

The stability of the biosphere was due to Gaia, not *homo sapiens*. Earth's vast and interconnected total genetic pool made the loss of even two and a half million species apparently of minimal practical significance. We can, however, take credit for the second great constancy: Basic materials did not, and have not, come into short supply. And it is humanity's own creations that must be thanked for this.

The first attempts to make robots, in the latter half of the previous century, were discouraging. Perhaps this was because, following popular fiction, these computer-controlled devices were perceived as *servants,* and expected to perform such tasks as house cleaning and maintenance. It took a long time to realize that compared to the uniformity of a mine, a water filtration plant, or an ocean floor, a human household is a vastly complex operating environment, intolerant of less than optimum performance. Beginning with the remotely operated deep-sea submersibles of the 1980s, robots in the 1990s and early 2000s began to be employed in every structured environment that was difficult and dangerous for a human, and in every situation where tasks could be clearly defined: mining for low-yield metals, for low-sulfur coals, and for iron ore on land, and for manganese, in the deep sea. Smart robots also made viable the mining of ores far poorer than those traditionally of economic value. Real material shortages are still a problem for the future, but that future will be in the twenty-second century.

Increased use of specialized robots has also decreased failure levels. Robot coal miners, for example, can sense ambient levels of methane directly and continuously, so that underground explosions have become a part of mining's primitive past. Miniaturized miners, a few micrometers across, monitor the sulfur levels in coal directly, and separate it out, while smart sensors in the support beams of the mines report continuously on stress and movement levels.

Even smaller (molecular sized) specialized robots remove toxins from our air and water at efficiency levels undreamed of in the twentieth-century scrubbers and filtration units. Since production plants and transportation systems are now obliged to use at least 90 percent of their air and water effluents, recycling has become close to perfect.

As our unpaid and unsleeping robots have decreased the cost of raw materials, so also have they decreased the costs of refining and manufacturing. This, together with the development of increasingly robust and versatile plastics and ceramics, has lowered the final cost for the consumer; products are far cheaper than they were half a century ago. Since the early 2000s, robots have replaced humans in most manufacturing, and in the agriculture of staple products, where high levels of judgment are not required. The danger and tedium of twentieth-century factory or agricultural duties would be deemed totally intolerable by today's workers.

Robots, and robot control of operations, today are almost everywhere. The general-purpose household robot, flexible in tasking and safe in all circumstances, has proven vastly difficult to develop. Robots smart enough to stimulate and interact with human activities over a broad range have been developed only in the past 10 years.

Humans can also take credit for the third great change of the first half of this century: widespread starvation, for so long the specter looking over our shoulder, has left the scene.

Starvation was banished as a result of three other changes. First, using recombinant DNA techniques, scientists build superplants that can thrive with high productivity in areas too saline, too cool, and too arid for earlier natural plants. Where only 15 of the 150 million square kilometers of the world's land were cultivated 50 years ago, and three-quarters of the whole judged unusable, today over 40 million square kilometers are cultivated, and the water demand for irrigation has been more than halved.

Second, the production of wholly artificial food, dreamed of for centuries, became a reality in the 2020s. It has never been wholly accepted or generally popular, and individuals claim to know the difference between real and synthetic foods (although double-blind taste tests show that they are deceiving themselves). But natural foodstuffs are preferred, and some individuals even grow some of their own food and make their own wine, although these cost vastly more than purchased products. However, artificially produced food stands as a bulwark against true shortages, and is available everywhere in time of need.

These two developments, artificial food and superplants, however, would have been meaningless had human population continued its blind and insensate increase. That it did not is due to one simple biological advance, already on the horizon half a century ago. This was flexible and foolproof contraception, and associated fertility control.

Although religious debate did not end in the 1990s (and has not even today), contraceptive pills, direct descendants of the antiprogestin pill of the 1990s, had by 2010 become cheap, safe, and ubiquitous. This put the choice of family size, firmly and finally, into the hands of the people who bore the children: the women. In a single generation, families of more than four children became the exception. By 2020, two children per family was, as it is now, the norm.

Today, the world's population of nine billion is even showing signs of a downturn. For the first time in two and a half centuries, ever since Malthus made his ominous prediction, the diapason of population growth is not sounding through the whole of human affairs. The Right to Bear Children, like the Right to Bear Arms, is an ongoing argument, one that has become less bitter since the average family size has decreased. The future no longer holds out to us a prospect of universal malnutrition.

Life at its most primitive is very simple: a sufficiency of food, clothing, and shelter. Once these are satisfied our demands become more complex. We begin to ask for more.

High on that list of increased demands comes our health. A large number of life's joys and miseries depend on how we feel. And when we do not feel good, we have come to expect that to change.

As the first half of the twenty-first century nears its close, we have become accustomed to noninvasive medical diagnosis, and to minimally invasive medical treatment. We expect external imaging sensors and internal "insensible" sensors to tell us what is wrong with us. We increasingly expect drugs to cure us. If we now consider it an inconvenience to swallow a pill-sized object, one containing its own sensors and capable of being guided internally to any point in the body without being felt by the patient, we should faint at the idea of the old discomforts and dreads: the drawing of blood, the catheters, the proctoscopies, the biopsies. Death, or even cure, used to be accompanied by a thousand indignities. That is no longer considered tolerable.

And drugs are able, more and more, to cure us. All immunological functions are now understood, one by-product of the genome mapping that was well under way half a century ago. Cancer, the "big problem" of former times, is an immunological deficiency disease that is now completely curable; just as important, so are a hundred other ailments, as "insignificant" (except to the victim) as asthma, hayfever, and hyperallergic reaction to everything from foods to dust.

The growth process for nerve, organ, and muscle cells is also understood. This has ended paraplegia, and made possible the replacement of eyes, limbs, and internal organs, besides breaking the curse of being freakishly taller, shorter, fatter, or thinner.

Another important factor that contributes to the enjoyment of life is leisure: The shrinking need for human labor as a result of widespread automation was predicted in the 1950s, but its social implications were misread. People foresaw massive unemployment. Instead, we have moved to today's ten-hour work week, with positions shared by ten or more individuals who are at work consecutively through the week. Although there is thus seven days a week service for everything, it is a rare individual who works for more than two of those days, and then it is by choice. The move to the two-day work week, plus vastly improved and widespread electronic communications and the freedom to work from one's home, has also made the words "rush hour" as much an anachronism as "computer error." The Robotic Revolution has proved to be rather like the Industrial Revolution, which though it forced lives of appalling toil on the less fortunate of the industrial nations at the outset, ultimately provided their descendants with vastly increased leisure and freedom.

The story of the improved quality of life of the past 50 years would be incomplete without reference to the phrases "civilian murder" and "global war." The former dwindled to its present negligible level when a basic distinction was drawn between the right to *bear* arms, and the right to *manufacture arms.* The first still exists, but strict curtailment of the second has led to a far safer world. Access to weapons able to kill many people quickly is now tightly controlled.

As for global war, this was already declining as a central concern by the 1990s. It finally vanished as a direct consequence of the world's increasing economic interdependence, with regional specialization on particular products. As people became unwilling to endure the loss of available products, the reluctance to cut their supply increased. We now take for granted our year-round access to produce of all nations, just as we take for granted our seven-day a week, 24 hour a day access to services.

Certainly, we do not live in paradise. There is still aggression, there are some hold-out diseases. And finally, there is still death. Human maximal life expectancy has not increased, even though we have come to expect a healthy old age, with a thousand times as many *vigorous* centenarians as half a century ago. Compared with every earlier age, the world today, for the average human, *is* Paradise.

Some of our still-sought changes are elusive, and may never happen. The world remains a Tower of Babel, with all attempts to create a universal language unsuccessful. Only the fact that 85 percent of the world's people have a working knowledge of English makes that less of a worry.

Universal literacy still eludes us. Worse than that, we are divided more than ever into the *can reads* and the *can't reads,* a division more devastating to the latter group than anything of race or heredity.

The long-promised cornucopia of space still withholds its bounty. Three-quarters of a century ago it was seen as the place where solar energy would be generated and beamed as microwaves to the Earth. Now we realize the historical inevitability of increased "energy density," which has been a theme through all human history as animal, human, and wind power gave way to water, steam, chemical, and finally nuclear power. Humanity will never return to a major dependence on the diffuse natural fusion energy of the sun, now that controlled and economical man-made fusion energy is (at last!) a reality.

Half a century ago, space, like the oceans, was also seen as a sometime utopia for human habitation. Today it still holds that future prospect; but for everyday life, earth orbit is the place where we put the dirtiest and most dangerous of human activities. For above all else, space is an *insulator,* a barrier against toxins of all kinds. We have at last learned the lesson of the Middle Ages, the privy and the well should not drain into each other.

We are still struggling to evolve the "ideal" political system. The Individualist Movement is having the formative effect on this century that democracy and communism had on the nineteenth and twentieth, but is an unfinished story. We still yearn for a system which sets great controls on rash and harmful actions, but few controls on benevolent ones.

The central ideas of Individualism are possible only because computers have changed the old economic imperative of mass production. The guiding industrial principle, "Every One the Same," which favored the productions of millions of identical copies, gave way early in this century to the principle, "Every One Different." With computer control, there is no increased cost for tuning products to individual tastes, and bigger potential markets. We are still working out the social and political consequences of this new production world. For example, houses, though mass produced in central locations, are yet infinitely varied. But we have not been able to define the optimal city size. Cities, meanwhile, have been shrinking without formal controls, with more and more at about 30,000 people. Though tiny by twentieth-century standards, these cities offer services at unprecedented levels.

Not everything, however, looks like progress. The right to individual choices is more important today than ever before, and this century has seen the development of a new "Bill of Rights" which provides increased right to individual views, and more right to individual life-style. Like it or not, that right includes the freedom to abuse one's own body. If the life of man is still, as Thomas Hobbes proclaimed, nasty, brutish, and short, it is so more from *choice* than from necessity. We have not reached the end human frailty, and perhaps never will.

Finally, we have not reached the end of science and technology, or even the beginning of the end. The understanding of the uses of silicon, which promises to dominate the inorganic world as carbon dominates the organic world, now seems to be in its infancy. Nanotechnology, with molecule-sized machines employed everywhere from inside of our own bodies to the deep interior of the Earth and the farthest reaches of space, has yet to achieve its perceived potential. The "simple" process of cellular differentiation remains a puzzle. The subnuclear world continues to produce as many surprises as it did half a century ago. And the structure of the cosmos, and its origin, is still a deep mystery.

We have much to learn. And we expect that much will change in the next 50 years. The words of George Santayana, writer nearly a century and half ago, seem as true today as in 1905: "Those who cannot remember the past are condemned to repeat it." Today, however, as a warning to us against complacency and rigid thinking we add their converse: "If you remember the past too well, you will see no way that the future can ever be different."

TWELVE

AN ADDRESS TO THE COUNCIL

Ben Bova

My fellow citizens of the world:

It is both an honor and a grave responsibility to assume the chairmanship of this newly formed World Council. The burdens that face us are immense. Our resources seem barely adequate to deal with the massive dislocations forced upon the world's people by population growth and climate shift.

Famine stalks much of the Southern Hemisphere. Even in the industrialized nations, life expectancies are declining. Civilization stands on the brink of a precipice, in danger of a fall from which it may never recover.

If I may have your indulgence, I want to review the more significant problems that beset us. I want to demonstrate how the interrelated nature of these problems has created a negative synergy that actually makes their totality much more difficult to solve than they would be as individual predicaments.

First, and foremost, is the continued explosive growth of world population. According to this morning's computer monitoring data, 10.7 billion human beings occupy Planet Earth. More than 60 percent of them live in urban areas. Overpopulation is outstripping the transportation facilities, water supplies, and food and waste removal systems of cities across the world. From New York City to Sáo Paulo, from Cairo to Tokyo, the world's cities have become festering ghettos rife with crime, drugs, and despair.

By today's end, another three hundred thousand babies will have been born. Each of these human beings will require food, shelter, education, and means of self-support. Each of these factors, in turn, will make demands on the planet's natural resources and energy. As long as the world's population continues to grow unchecked, all our efforts to increase the Earth's productivity will continue to be swallowed up by the increasingly numbers of hungry mouths.

Exacerbating the population problem is the problem of climate shift: as if the punishment for the sins of our fathers is being visited upon us. The pollution wrought by two and a half centuries of industrialization has pushed global temperatures into true greenhouse levels. Lands that were once fertile are turning into deserts. Sea levels are rising all around the world, threatening hundreds of millions of families with inundation, which will force them to relocate. Annual monsoons and tropical storms have increased in ferocity, as anyone who lives in a coastal area knows.

The seven-decade long Petroleum Wars have ended, thanks largely to the dedication of the International Peacekeeping Force—although perhaps even larger thanks should go to the scientists and engineers who brought us practical and efficient nuclear fusion power.

To almost everyone's astonishment, including many here this morning, the International Peacekeeping Force has worked. The peacekeepers did not prevent aggressors from attacking their neighbors—at first. But the national leaders slowly learned that any attack by any nation upon any other nation would be met swiftly by powerful defensive forces, in the name of the united human race. Gradually the IPF sent a clear message to would-be aggressors that the price for military assault was higher than any possible reward might be.

The peacekeepers can serve as a model of how we can work together for the betterment of the human race. They have virtually eliminated the scourge of war: we must work together to eliminate the potential causes of future strife.

The Petroleum Wars, of course, were not entirely military in character: Economic warfare, political maneuvering, even public-relations tactics, were all part of the 70-year struggle. Much like the old Cold War between the North Atlantic nations and the former Soviet Union, the Petroleum Wars were fought at many levels and exacted a terrible toll in human life, squandered resources, and environmental degradation.

As I said, today we stand at the edge of a precipice. The human race has reached a turning point, one of those moments in history where the decisions we make now will be the fate of human kind for centuries to come—perhaps forever. The fundamental problem that faces us is nothing less than the choice between the survival of civilization or its extinction.

Technologically, our civilization has achieved great things. We can fly through space and build habitats at the bottom of the ocean. Our medical sciences have extended life expectancy to the point where difficult legal and ethical questions are being argued over the state's right to impose a limit on human life spans, as opposed to the individual's right not only to live for many more decades than a century, but to have one's body preserved at the point of death in the hopes of being revived later, cured of the "fatal" disease, and then resume living.

But these wonderful achievements have been restricted to the rich. The overwhelming majority of the world's peoples are poor. They live and mate and die as they always have; Their numbers are growing exponentially. With each generation they grow poorer, both in absolute terms and in relation to the rich. The rich, meanwhile, control their family sizes;

the wealthy nations have stabilized their populations and erected virtually impassable barriers against immigration.

How long can this planet continue to exist with a steadily growing population of extremely poor people and a small, stable population of extremely wealthy people? How long can one-quarter of the world's people consume 90 percent of its natural resources, while the other three-quarters eke out a precarious living in squalor, misery and frustration?

As a great sage said over two thousand years ago, "A house divided against itself cannot stand."

The Petroleum Wars were a symptom of this fundamental problem. Behind the politics and the military combat was the desperation of poor people struggling to obtain some scant slice of the world's riches. The wars have ended, but the underlying problem remains. If we fail to solve it, new fighting will erupt: revolutions and terrorism that even the peacekeepers will be helpless against.

Let me emphasize this point: Unless we find the means to alleviate the poverty and hunger of three-quarters of the world's population, our civilization will crumble and collapse into a new Dark Age of incessant warfare and chaos. The rich will be swept under by the growing tide of the desperately poor. The world's population problem will be solved by the Four Horsemen of antiquity: Famine, Disease, War and Death.

We have the tools, the knowledge, the technology and the understanding to build a new world society that is fair, and free, and flourishing. But do we have the will, the courage, to create fundamental changes in the world's existing political and economic structures?

How do we narrow the vast and growing gap between the rich and the poor? How can we avert the famines and end the poverty that already hold in their pitiless grasp three out of every four human beings on Earth?

I can see two possible approaches: coercion or cooperation. I much prefer cooperation. But let me say a few words about coercion.

There are two possible ways to use coercive tactics to increase the wealth of the poor nations: 1) force the rich nations to give up enough of their wealth so that the poor nations may advance economically, and 2) force the poor nations to limit population growth and adjust their economies for long-term growth rather than the stopgap measures they now adopt on an as-needed basis.

I think you can see the difficulties with each of these tactics. The rich nations will resist a drastic redistribution of wealth. They will fight such measures politically, economically, and, failing all else, militarily. If all—or even most—industrialized nations took up arms against our World Council, not even the IPF could prevent them from sweeping us into the dustbin of history.

Coercing the poor may then seem easier. But how do we enforce population limits for entire continents? Do we have the moral superiority to tell billions of people that they must ignore the dictates of their religions and their social customs in favor of *our* vision of what the world should be?

The poor will resist such attempts. They may not battle us militarily, but simply ignore us and continue to have babies. What would we do then: slaughter the innocents in the name of global economic progress?

My friends, coercion will not work. Not only is it wrong, it is ineffective.

That leaves us with no other choice except cooperation.

A wise man once observed that the "problem with the world is not that there are so many poor people; it's that there are not enough rich people." The difference is subtle, perhaps, but very real. We cannot force the rich to give their wealth to the poor.

The alternative, then is to somehow make the poor rich *without* pauperizing the rich.

Some of you who make up this council are scientists and engineers. Most of you are not. To all of you, I say that technology—toolmaking—is the way human beings adapt to their environment. We do not grow wings, we build aeroplanes. We do not have the muscular strength of the gorilla or the fleetness of the antelope, yet we lift tons of weight at the touch of a button and race across land and sea faster than any gazelle or dolphin.

Technology, however, is not, and should not be, an end in itself. Toolmaking by itself cannot solve our problems; tool *using*, instead, has redeemed humanity time and again, since the Promethean days when man first tamed fire. Our task is to use the bright, shining tools that the technologist have produced to better the economic and social conditions of man.

Let me give you an example of how some of our most sophisticated tools are being poorly used.

The work force in industrialized nations consists of almost as many robots as human workers. This is especially true in the manufacturing and extractive industries, where robots "man" factories, mines, and farms. A generation ago, when truly useful and adaptable robots began to enter the work force in Europe and North America, there was a great surge of labor unrest: People feared being replaced by robots.

It was the Japanese who showed the way around this problem. In one form or another, human employees formed partnerships both with their robot coworkers and with their employees. They earned incomes from the wealth generated by the robots who labored for them. This system of employee ownership has allowed increasingly sophisticated robots to enter the work force of the industrialized nations smoothly, with little labor strife and maximum profitability for all. But the increasingly roboticized work force of the industrialized nations has had devastating effects on the economies of the poorer nations.

Robots in England, for example, produce clothing more cheaply than the lowest-paid workers in Angola or Bangladesh. Robots in California do the "stoop labor" of harvesting that was formerly done by migrant Mexican farm workers. California grows richer; Mexico grows poorer. Robots are widening the gap between the rich and the poor. Indeed, robots are even replacing domestic servants among the very rich.

Can robots help to make the poor nations wealthier? Can, for example, robots turn a squalid farming village in Guatemala into a thriving and prosperous community? No. Not by themselves. The people of that village are unprepared for such a leap into the modern world. They lack the education and the social framework to deal with machines that move and work and think. Their village, their entire nation, lacks the economic foundation to employ robot labor usefully.

But robots can indirectly help the poor nations by helping to produce the wealth needed to start those nations on the road to riches. It is inevitable that the rich nations must devote some portion of their wealth for the salvation of the poor

I propose that we, the World Council, impose a tax upon all nations, based on the ratio of each individual nation's gross national product (GNP) in comparison to the mean GNP of all the nations. Thus, the very richest nations would pay the largest tax, while the very poorest nations would have a negative tax; they would receive income from the tax fund.

The tax income thus received by a nation could then be devoted to long-term programs to improve its economy. Thus the rich nations would pay to make the poor nations richer. In essence, the roboticized work force of the rich nations would help increase the wealth of the poor.

This proposal smacks of the coercion I spoke of just a moment ago. But I ask the leaders of the rich nations to undertake this small sacrifice willingly. I have no intention of coercing any nation, and this World Council certainly does not have the power to do so. We need your cooperation, rich and poor alike.

I am not merely proposing new taxes, however. If all we do is shift wealth from the rich to the poor, our efforts will be resisted and ultimately fail. We must therefore develop, wherever possible, the means and the opportunity to generate new wealth. Our primary goal should be to enlarge the real wealth of the human race.

The places to find new resources are the frontiers of our existing habitat. Obviously, our physical frontiers are the world's deep oceans and the even greater depths of outer space. However, there also are mental frontiers. And the research laboratories are frontier country.

Perhaps the best news of this century is that a safe, efficient nuclear fusion power system has finally been developed to the point where it can deliver reliable electrical energy. After nearly a century of research, our scientists and engineers have harnessed the energy of the sun and the stars. Our everlasting thanks to the perseverance of the brilliant men and women who have given us this inestimable gift. In my mind, it ranks with the original gift of fire, back in the mists of prehistory

The fusion process is so energetic that there is enough fusion fuel in an 8-ounce glass of water to equal the energy content of five hundred thousand barrels of petroleum! And less than one percent of the water is consumed! The rest is available for drinking, irrigation, or other uses.

I dwell on the prospects of fusion power because it is the key to the ultimate solution of the tremendous problems we face. With clean, efficient, and ultimately cheap fusion power we could bring reliable energy to the poorest of nations. In areas where the groundwater supplies have either disappeared or become contaminated, we will have the energy to desalinate seawater economically. This water could then be pumped over long distances to counteract desertification.

Another important use for fusion energy will be in the recycling of waste materials. For more than 50 years we have looked to recycling metals and plastics as a means of cleaning the environment. But recycling requires energy, which so far has been prohibitively expensive. Fusion power will make recycling inexpensive and profitable. Moreover, we will have a new source for raw materials when fusion powered recycling centers reduce our waste materials into highly refined elements. Recycled metals and chemicals will, in fact, be cheaper, in many instances, than digging new ones.

There is another frontier that we need to consider. I speak of the tremendous opportunities and problems created by the biological sciences and the concomitant biotechnologies.

Having delved into the very core of living cells, biologists have learned the secrets of our genes so well that it is possible to extend human life spans far beyond a century. There are even glimmers of literal immortality in the latest research reports. Although with the aforementioned population problems, we have scant need for longer life spans; yet, who would reject such an opportunity?

Of more immediate import is the enormous impact that biotechnology is making in agriculture and medicine. New strains of food crops genetically engineered to withstand drought, heat, frost, pests, or other hostile conditions, can greatly increase the world's food supply. Genetically engineered bacteria now "fix" nitrogen in many food crops, eliminating or greatly reducing the need for artificial fertilizers.

These powerful aids to our efforts to feed the world, however, are not entirely without risks. The temptation to plant nothing but these new "supercrops" could cause havoc if some factor were overlooked and the entire crop failed. Diversification must remain the watchword.

Biotechnology, like other technologies, is a double-edged sword. The very breakthroughs that provide cures for genetic diseases such as diabetes and cystic fibrosis could well become agents of unparalleled virulence. I am not suggesting a Frankenstein scenario, but I do insist that biotechnology laboratories be continually monitored by World Council Agencies. Our aim must be to reap the benefits while minimizing the risks

Now to the physical frontiers. We know that the world's oceans hold immense resources of food, energy, and raw materials. Today, we are consciously reproducing in the oceans the Neolithic Revolution that our ancestors produced on land some ten thousand years ago: We are moving from merely gathering food from the sea to deliberately growing food there. Fish farms, algae farms, carefully tended beds for shellfish—these and more can eventually produce far more food, per hectare of sea surface and per calorie of energy input, than farms on land.

There is energy in the sea, as well. By tapping the temperature differences between the cold deep layers of the ocean and the sun-warmed surface layers, it is possible to produce abundant electrical power without harming the environment. Thus, industrial facilities and human habitats on the sea can be self-sufficient in energy, and could even sell energy to consumers on land.

Seabed mining is already yielding important metals such as magnesium, manganese, copper, and molybdenum. Future efforts can only enhance these resources, besides also increasing humankind's supply of wealth.

Today, three Solar Power Satellites provide half the electrical energy used by Japan. As wide as Manhattan Island, placed in high orbits where sunlight constantly drenches their broad panels of solar cells, the satellites beam energy in the form of microwaves to receiving "antenna farms" built off the coasts of Honshu and Kyushu. A multinational consortium is attempting to raise capital for two more Solar Power Satellites, one each for Europe and North America. Furthermore, Moonbase Inc., recently announced plans to build solar-power "farms" on the Moon's surface and offer the electrical energy they produce to consumers either on Earth or to the growing number of manufacturing and research stations in orbits between Earth and the Moon.

The old dream of mining the asteroids for their metals and other raw materials has not yet been realized. Asteroids, for the most part, are twice the distance from Earth as Mars, and there is as yet no economic necessity to go that far for metals that can be obtained on Earth.

Outer space is a harsh and dangerous frontier. It will never serve as an outlet for the Earth's growing population. There will not be mass emigrations to space—not in any foreseeable future. However, space *is* rich in energy and raw materials. And it offers unique environments in which inventive humans can produce goods and services that are impossible to produce on Earth.

The highest-quality metal alloys are manufactured in zero-gravity orbital facilities, as are the best crystals for our electronics industry. Much of the pharmaceutical manufacturing is taking place in space facilities, where, besides zero gravity, there is abundant solar energy and the ultraclean environment of high vacuum. Also, a small but lucrative tourist industry has sprung up in Earth orbit—for those few rich enough to afford it.

Outer space, the deep oceans, the research laboratories around the world: These are the frontiers from which we may generate the new wealth that can alleviate the crises of the poor nations.

But all our efforts, hopes, and plans can come to naught if the climate continues to deteriorate. The terrifying facts are that every year, in this century, thousands of square kilometers of productive lands have turned into desert or useless scrub land. More thousands of square kilometers have been inundated by rising sea levels.

The global climate is warming, heated by greenhouse bases that we ourselves produce. For the first time in history, the actions of the human race are overshadowing the natural processes of climate and weather. We are overburdening the atmosphere, the oceans, and the land with our own filth. We must reverse the manmade trends that are altering our environment so swiftly that natural processes are being overwhelmed.

We have the tools to accomplish what must be done. Modern technology has given us cheap fusion energy, solar power, and energy from the oceans. All these systems are renewable and do not produce greenhouse gases. Superconducting motors and batteries, which can replace existing petroleum-burning engines, can end the smog problem that plagues the world's major cities.

Our biotechnologies offer the opportunity of using biological means of increasing farm productivity and controlling pests, rather than brute-force methods of artificial fertilizers and pesticides.

We have the tools. Do we have the intelligence and the courage to use them wisely? Military men speak of "friction," meaning the thousand-and-one individual misunderstandings and resistances that arise in the heat of battle. "Friction" is what comes between the general's brilliant plan and the actual outcome of the combat. You and I face "friction," also.

The sad fact is that human beings change their attitudes only slowly.

A villager who owns a petroleum-powered tractor has little incentive to incur debt for a new electrically powered one, especially if he must learn to deal with superconducting machinery that requires liquid nitrogen.

A corporate executive who oversees a petrochemical plant is not going to endorse a shift to biotechnology that will replace everything he knows and leave him feeling useless.

A factory worker who controls several robots will not endorse a tax increase earmarked for the benefit of peoples of different lands.

And, saddest of all, there are still politicians who think nothing of siphoning funds from aid projects for their own personal use. How many earlier attempts to assist the poor have merely served to make certain political leaders rich?

This is the kind of friction we must overcome. As I said earlier, we stand at the edge of a precipice. All of us, rich and poor alike, are staring at the extinction of our civilization. I fear that the cumulative effects of climate shift, massively crowded cities, and growing hunger and poverty will combine to tear apart the very fabric of society. Chaos and bloodshed, such as the world has never seen, will sweep across the face of the Earth.

That is the nightmare we must avoid. Our policy must point toward the dream of a free, fair, and flourishing world society, where we fight with all our strength and all our wisdom against humankind's ancient and remorseless enemies: poverty, hunger, ignorance, and despair.

It is far from certain that we can win this battle. But we must try, for if we do not, then the end of civilization is at hand and the legacy we give to our children will be endless savagery and pain.

THIRTEEN

A NIGHTMARE

Morton A. Kaplan

"Nightmare" is an appropriate designation for the scenario that follows. Although it will read like an account of a frightening dream, the world of 2044 might resemble that dream if we extrapolate certain tendencies in society.

The seeds of this nightmare, bizarre though they are, lie in the real world. And the arguments for many of the individual steps that might bring it to pass in some cases might even be considered praiseworthy.

Much of what occurs in this scenario is not original. Aldous Huxley in *Brave New World* and George Orwell in *1984,* as well as many other novelists, created anti-utopias in which omnipresent social controls radically changed the nature of man and of society. The anti-utopias usually have been regarded as merely imaginative extrapolations that were technologically infeasible. Unfortunately, a few short years later, as we now write, it is less difficult to extrapolate from existing progress in the sciences to the technological developments necessary to produce these anti-utopias. Huxley's Alphas and Betas would still seem scientific marvels; yet recent advances in genetics, with specific reference to the unraveling of the DNA code, show them to be within the range of possibility. Many of the scientific developments of the future could be even more consequential than nuclear fission, less controllable by current techniques, and more inconsistent with existing social institutions and cultural values.

Many of the demands for increased social controls will stem from the vulnerabilities of modern complex societies. Although modern complex society has both greater instantaneous and long-range flexibility than simpler societies, it also has less redundancy and more bottlenecks that could affect the whole society. Thus major interruptions that overwhelm its instantaneous or short-run adjustment capability, or occur too suddenly for its long-range flexibility to get a chance to work, might cause great damage to the society. Because simple societies are less interdependent than modern complex societies, their breakdowns are often much less total in their effects. The modern industrial society is highly differentiated and therefore requires greater integration in order to function effectively. The disrupted complex society, under at least some important conditions, might not be able to sustain even the low level of productivity that is normal to a simple society.

The greater wealth and improved technology of modern society provide us with many important advantages and freedoms. Diversity of life-style, despite mass production and mass man, is possible today in a way that in the past was available only to a few of the elite or of the wealthy. The great diversity of modern society, however, requires a geometric increase in the organization of modern life. We become increasingly sensitive to the disturbances produced by others, particularly with respect to the environment. And this sensitivity, in turn, requires greater and greater social control in order to maintain the peace and stability of the system. One need not assume the triumph of the police mentality, or the intrusion of motivations denigrative of human dignity, to foresee that many restrictions on human liberty will have valid and attractive rationales, even rationales related to the liberty of one's fellows. Federal safety regulations for automobile manufacturers and tests for drivers increase the "freedom" of the license-holding driver to drive in safety. Coercive treatment for the mentally ill raises the probability that they will be able to lead freely constructive lives. Therapeutic abortions, through the death of the fetus, increase the freedom of the mother. And the biological adaptation of man to his ecological niche in an extremely complicated and overpopulated society will increase his freedom to live a satisfying and useful life.

We have, of course, omitted the crucial qualification: under the new developing conditions. It is still possible that the terminus of the process would be inconsistent with anything we would regard as freedom or dignity or even human. The evolution of society might produce the devolution of man. The adaptability (and superiority) of man has, heretofore, consisted in his lack of specialized adaption (unlike the lesser animals).

In the not-too-distant future, we might be adapted in a specialized sense, while society, through the control of genetic science, maintains its general adaptability by fitting us to the various tasks that time and environment provide. The survival of the fittest may be replaced by the fitting of the survivors.

The nightmare is bizarre, but consider a few of the factors that could facilitate and seem to justify a controlled society, or a few that could make it feasible, although the two categories are neither necessarily nor entirely distinct. Overpopulation and organizational complexity have already been mentioned. How can China survive with two billion people and India with one billion—figures that are no longer preposterous?

The greater susceptibility of society in general to disruption could create opportunities for deliberate intrasocietal attacks accompanied by blackmail, to say nothing of organized crime on a novel scale. When nuclear weapons become subject to criminal access (miniaturization will help to bring this about), and when criminal or political conspiracies become capable of bringing civil government to a halt through the disruption of the computerized networks upon which it will depend, the only alternative to a new feudalism (without the mitigating social features of the old) might be forms of surveillance and control far surpassing any now in existence.

Access to places of amusement and museums might have to be rationed, food substitutes developed, access to new forms of socialized housing (mile-high community units) regulated, and scarce medical facilities (replacement organs, esoteric remedies, very skilled surgeons, and so forth) allocated. Some humans might have to be adjusted to environments different from the earth's surface.

Clearly it could be erroneous to compare man to the lemming, but every known animal species on which the experiment has been carried out is dysfunctionally disorganized by overcrowding. Rules governing mutual adjustment (whether social or legal) wili necessarily be very stringent. There will be a strong emphasis on adjustment (other-directed orientations) in place of individualism. Resort to drugs, otherworldly religions, delinquency, crime, and mental disease (as a way of "acting out") could increase significantly, requiring medical, social, and criminal sanctions to prevent or to contain those forms of disturbance that are excessively dysfunctional for the social and political systems.

The consequences of dislocations and of mistakes in the production, distribution, and control functions of business and government are likely to be so huge that facilities for coping with them must take precedence over civil liberties or private pursuits and property. The blackout of the eastern United States in 1965 only suggests what can go wrong in the future. (That particular disturbance need not have occurred with proper systems design, including cutoffs and redundancy, with their concomitant costs—but that is the problem. On the one hand, there could be extremely sophisticated and prudent systems designs or, on the other hand, there could be extensive control and supervision to avoid the possibility or consequences of relatively farfetched disasters.) Needs for control and surveillance will likely develop to utilize the technological capabilities that are present in the system. Technological developments, in addition to meeting environmental requirements, will likely produce needs to satisfy the technological capabilities.

It is already possible to monitor conversations by the disturbances they produce on window panes and to photograph documents through windows at great distances. Television monitors both indoors and out might become common as the techniques become cheaper. Voices and faces, and perhaps retinal patterns, might be checked immediately by advanced computers working through nationwide banks of identifiers. Quite apart from credit needs, the means to maintain continuous checks on the entire population and automatically to scan them for disturbing words or phrases will likely be available by the year 2044. It might occur that only those with enormous resources will be able—and in even these cases only partly and perhaps only through bribery or by political manipulation—to avoid some monitoring or to interfere with transmission of the data. At the minimum, if the monitoring exists, new code languages will develop in efforts to evade some of the consequences.

There are cases today of individuals who are kept biologically alive, in an attempt to avoid testamentary consequences, by bizarre and uncomfortable medical techniques long past the point at which the physician, the individual, or the family would otherwise prefer a natural death. As facilities for replacing human parts increase, including artificial stimulation or substitution for certain brain functions, court cases will almost inevitably arise over the issue of when a man ceases to be himself. (Consider the case of a man who has had most of his intellectual capabilities replaced by a computer or even by a transplanted brain.) Ultimate resort to these techniques would come under government regulation. If overpopulation and means for increasing longevity increase to some as yet unspecified limit, the right to bear children and to resort to longevity techniques could be controlled by the government. How these issues, which are so politically potent, could be handled presents a challenge to the imagination. If such issues arise, and they well might, then the consequences for what we now regard as civilized human standards are obviously enormous.

It is not unlikely that there will become available by the year 2044 drugs and other behavior controls capable of producing personality changes at will, of rewarding activities by hormonal flows (perhaps by remote control) that overcome rational ego or super-ego objections to continuation of the activities, and of punishing other activities. Alternative techniques include radio waves, ultrasonic impulses that cause uneasiness, induced hallucinations, and various forms of educative devices operating from infancy. These might be so effective that continuous control techniques would be superfluous, although available for obdurate cases. Much of this might be available or imposed under the rubric of mental hygiene, simply because such intrusions on individual freedom are not likely to occur except for highly persuasive reasons.

It is not difficult for an American to understand that a dictatorship—even a benevolent one—would use such techniques. The Soviet Union, under its communist dictators, sent some of its important literary figures to mental institutions; it is well to remember that the United States sent Ezra Pound to such an institution, although it did so as an act of kindness. It is difficult to accept that such techniques would be used widely in the United States until we recall the extent to which they are already legitimized. Hundreds of psychiatrists were apparently willing in 1964 to lend their names to the conclusion that Barry Goldwater was mentally unsound although they had not examined him. Our culture is attuned to the concept of mental illness and its cure: the modern concept is to rehabilitate rather than to punish criminals because of the belief that crime results from mental illness.

Delinquents are guided by social workers. Disturbed schoolchildren are treated by guidance counselors. Parents read psychology to learn how to raise their children. The rhetoric of our time—and most of it is quite genuine and functional—is the rhetoric of mental adjustment and treatment. Our national pastime is self-medication with tranquilizers and with other drugs that affect the psychological condition of the individual. Rather than doubting that Americans would use the most advanced techniques that become available—and as systematically as possible—there is reason to doubt that there would be much effective resistance. We do not need the rationale of a political ideology to justify control of the masses. We have our own myth of adjustment and of mental balance; antisocial behavior, as interpreted by the received truth of the day, is sufficient to indicate the desirability of treatment. In the 1960s, some of the New Left—self-proclaimed rebels against social conformity—advocated (perhaps seriously, perhaps not) placing LSD into dormitory food to free the mass of students from their "false"—hence also "sick"—beliefs. Even those of us who criticized this abuse of terminology at the time often thought of some of the New Left or of other social rebels as "merely sick."

As antisocial behavior becomes less tolerable as a result of the increasing complexity and crowding of society, are we not likely to treat what we cannot tolerate? No doubt, if we were ever presented with this kind of future as a direct and systematic alternative to the present, we would not opt for it. But we are not presented with choices in this fashion. Each adaptation that helps partly to produce the future is considered on the basis of incremental costs and benefits, of marginal changes; it is possible that each incremental change or marginal benefit might seem to outweigh the costs, that the benefits will be clear and the transition to the nightmarish future hypothetical.

So far the nightmare is one that could conceivably be produced by the year 2044 or shortly thereafter. The ethical problems that could be caused when we learn to produce humans or variations of humans in the laboratory are unlikely to occur until well after the year 2044. Laboratory men who are indistinguishable from ordinary men, we hope, would be granted the rights of natural men. Specialized laboratory-created beings that differ from natural men, but that do possess the ability to reason, are more likely not to be granted full rights (a decision we would regard today as ethically monstrous). If they are granted human rights, questions might then be raised about the rights of adapted and specialized men who are indistinguishable from manufactured adapted and specialized men. In any event, the scope and variety of restrictions upon full natural men will be enormously increased.

Further problems would arise if bionic computers are made that perform many of the tasks of men and develop creative capabilities. As the distinction between man and lesser creatures and machines begins to shade off, the uniqueness of man and the rights attributed to this uniqueness may begin to attenuate. The vulnerability of the political system to shocks and to disruptions could reinforce arguments for the restriction of man based upon the substitutability of manufactured men and bionic machines. A creature that is superfluous as well as dangerous might appear difficult to defend. If the aesthetic function of man also degenerates, then the argument that man is a unique cultural being will also have attenuated. Bionic computers, for example, might be able to produce real art. Even now, some popular music (and even some classical records) are produced as much by equipment and by mechanical interventions in the performance as by "natural" performances with "normal" instruments. More than this, creative bionic computers might produce music that is genuinely more creative than some of the current musical fashions, for instance, aleatoric music.

If athletes begin to make use of prosthetic devices to improve their performances, we might gradually produce almost entirely mechanical athletes, for whom bionic robots might eventually substitute. As this process continues, man's confidence in himself and in his role might be seriously undercut. His vision of himself as a unique being, so essential apparently to his sense of identity, might be destroyed.

A variety of strange religions would spring up in efforts to explain this peculiar universe. Such religions might attempt to glorify man in ways that repudiate the rational and scientific interpretations that have flourished since the Renaissance; or they might be masochistic and denigrative of man. More likely both types would flourish under the suggested conditions.

Perhaps many (most?) men would be kept in a permanently drugged (pacified?) state and adapted to the ecology to which they are assigned according to some computerized calculation. The central government would so likely be swamped by the problem of keeping the system functioning properly that it would be concerned only with marginal and immediate problems rather than with the increasing repulsiveness of the entire system. In any event, there might be no rational or moral (whatever these terms may mean in such a bizarre mid-twenty-first century) feasible solution that does not reject modern technology or condemn billions of surplus humans to death or to deprivation. The twenty-first century would no more be able to return to the world of the twentieth century than we could return to the golden age of Greece.

Efforts to control the situation would doubtless occur—perhaps as desperate measures. For instance, the political and intellectual elite might distribute contraceptive drugs through the food supply that could be counteracted only by other drugs restricted to the elite. The rationalization could be persuasive: with automation, production would not be disrupted and population might drop to tenable limits that permit humane standards.

Yet the cure might be more brutalizing than the problem. In any event, this "cure" assumes that a technocracy, or oligarchy, or "aristocracy" controls the political system. As technological innovations are made and biological manufacture and reproduction intensify, the legitimacy that invests political democracy might deteriorate; the political bases from which these encroachments can be resisted might be undermined. If, at the same time, the system becomes so complex that it can be worked only from the vantage point of the memory banks of a centralized national or worldwide computer system, political and military capabilities would, in fact, be concentrated in one (if not monolithic, at least centralized) control center. Particularly if creative bionic computers and human-computer feedback circuits are involved in the central apparatus, control may pass from humans to machines, in which case, although population might be limited since it serves no useful function, humanity might be kept in a perpetually drugged and/or subservient state, to the extent to which it is permitted to persist. This would prevent rebellion and disturbance or other "undesirable" interferences with the maintenance of the system. By determining what information to feed back to the computer-linked controllers and by manipulating the logic of the problem, the computers might gradually gain control of the entire system. This might result, not from some analogous organic urge to control or even to destroy, although the possibility that this complex might enter a condition equivalent to madness can hardly be dismissed, but from an effort to reinforce stasis; the bionic central computers are likely to view humans as defective, both emotionally and logically.

Because of the enormous importance of the national computer networks for planning and control, they might become the focus of politics, conspiracy, and intraelite coups in the event they do not secure the kind of control intimated above. Quite possibly efforts by political groups to seize control of the central computers might themselves disrupt the computer functions, at least temporarily, and produce crises or disturbances that affect the operations of and prognosis for the system. Advanced weapons systems that operate on a computerized basis would make consensus among the population and support within the armed forces almost irrelevant. Some types of weapons could be individualized and could, in effect, "home in" on voice and sight patterns of particular individuals identified from the national population register as enemies. Others would be used against large groupings but would themselves not require any human agency other than the programming of the computer. These weapons would use advanced surveillance techniques to find and to destroy their targets or alternatively to incapacitate them. Again, the patterns they would home in on, whether group or individual, would be transmitted from a central registry. Pickup of prisoners would be automated.

The nightmare we have sketched is bizarre, implausible, inhumane, and evil. But it is not impossible. Incremental decisions could lead us from the present to such a nightmarish future. Even if this future were regarded as unlikely and improbable, it is worth considerable thought to reduce that likelihood and plausibility still more. Many unlikely and implausible things have occurred in the past. The nightmare we have sketched would place us beyond redemption.

Note

This chapter is updated from a chapter in *Macropolitics: Essays on the Philosophy and Science of Politics,* by Morton A. Kaplan (Manchester, N.H.: Irvington, 1968).

FOURTEEN

A LAND OF EMPTY ABUNDANCE

Jerry Pournelle

People often assume that science fiction writers can predict the future. Although some of us sometimes pretend to do that, the reality is that no one can predict the future. What "futurists" from Bertrand de Jouvenal (Futuribles) to pulp science fiction writers do is describe a plausible future: one they believe could be, or, at worst, one they can make you believe could be.

Such projections can be optimistic or pessimistic, and can take into account various trends—or ignore them. I choose the "cautiously optimistic" scenario as it fits well with my personal view of the world: Things do get better, sometimes a lot better, but for every several steps forward there is at least one step back, and a couple sideways.

With that in mind, let's examine where cautious optimism might take us in the year 2044.

One trend is clear, both from the projections of scientists, and from scientific literature: Technology improves by leaps and bounds, and material wealth flows in abundance. For much of the world, things are getting better. People live longer, eat better, and have more material possessions; and where they do not, the failure stems from political rather than technological reasons.

Indeed, it is clear to me that by 2044 we will certainly have the technical capability to supply every person on Earth with material possessions equivalent, at least, to what the average American enjoyed in, say, 1940; and this, without undue strain.

There is no energy shortage if we don't want there to be one. Despite the popular myth, no responsible official ever said that nuclear power plants would make electricity "too cheap to meter"; but it is true that Japan and France are greatly reducing the energy costs for their citizens by constructing nuclear power plants, and other countries are following suit. There are some dangers to nuclear power, but they are not insurmountable.

In addition to nuclear power plants, there are various forms of solar power. These include Ocean Thermal (OTEC), which has been successfully demonstrated off the Kona Coast of Hawaii; ground based solar; and Space Solar Power Satellites (SSPS). OTEC works well but can only be built in areas fairly remote from where the power is needed. Ground-based solar power is too diffuse and too irregular—not available on cloudy days, or at night. Space based solar power systems are quite feasible. For less than the cost of the Persian Gulf War, the United States could have built the first of a series of Space Solar Power Satellites; proved the concept; built the first ground stations; and, most importantly, built a fleet of new generation space transportation systems for lofting the required materials to orbit.

Of course SSPS provides only electrical energy, and most of our economy demands oil; but almost any stationary oil burning system can be converted to electricity. Many mobile systems would work as well with electricity as with oil—or better; others can be powered by propane.

SSPS has a further benefit. One way to reduce the cost of SSPS systems is to use extraterrestrial materials in their construction. It may or may not be possible to build the silicon or gallium arsenide solar electric conversion cells on the Moon, but it is certain that the Moon contains everything needed for the actual structure of the SSPS; thus one benefit of SSPS will be a permanent lunar colony.

(Sanity check: the Lunar Society already has the names of over 300 highly qualified and technically skilled Americans willing to live permanently on the Moon.)

Access to the Moon means access to all the solar system; and it is easily proved that 90 percent of all the material resources easily available to humanity is not on the Earth. The Moon contains most of the heavy elements—iron, titanium, silicon, aluminum—needed by heavy industry, as well as more than abundant supplies of oxygen. It lacks many of the lighter elements, such as carbon, chlorine, and hydrogen. These could be supplied from Earth; but it is more likely that they will be supplied from stations in the asteroid belt. The Moon, Mars, and the asteroids among them contain everything needed for an industrial economy; moreover, any raw materials lacking on Earth can be supplied to Earth.

Moreover, it is quite feasible to build colonies in space. The late Dandrige Cole once predicted that the majority of humanity would live on space colonies within 500 years. He may well have been right, assuming that the human race continues to exist for that long.

Some projections of the future raise questions about food production and pollution: will we starve, or so destroy the environment that we cannot live in it?

It seems unlikely. Food and pollution are not primary problems: they are energy problems. Given sufficient energy we can produce as much food as we like, if need be, by high intensity means such as hydroponics and greenhouses. Pollution is similar: given enough energy, pollutants can be transformed into manageable products; if need be, disassembled into their constituent elements.

It's fun to go into the details, but in fact there's no need. On any trend analysis whatever, there is no real danger that the world in 2044 will be short of the technology required to produce abundant material goods at reasonable cost, both economic and environmental. We have the science and technology necessary so that by 2044 we can, if we will, provide luxury for many, plenty for the majority, and enough for all mankind.

Why, then, is our optimism cautious? Why do we feel uneasy when contemplating the future? The Cold War has ended. Technology pours forth. When we contemplate food, energy., material goods, and even pollution, we have good reason for optimism, yet, most of us are not optimists; for we see some frightening trends.

First, political: The United States is governed by a Congress which has less turnover than the British House of Lords. Our politicians pay themselves and their aides more than most of us ever expect to make, while simultaneously exempting themselves from the laws they pass for us to live by. The result is that "We the people" no longer think of "our government"; rather, we see "the government." Cynicism abounds.

Meanwhile, our economy gets more and more bogged down in rules and regulations.

Example: After years of study the national commission on acid rain reported that industrial pollution isn't the problem; that most of the acid rain isn't a serious problem anyway; that some lakes are threatened, largely by the runoff from forests that are not allowed to burn naturally (thus converting the tannins and other acidic forest products to smoke); but those lakes can be saved by a few tens of thousands of dollars spent on lime each year.

But instead of following the scientific advice, we have brought in a monster political program, that will add many billions of dollars to the costs of industry, and that will accomplish almost nothing toward ending lake acidity. Few who have taken the trouble to inform themselves of the acid rain situation want the acid rain bill; but the monster has political life of its own now, and it seems that nothing can be done.

Example: After years of discussion about Global Warming, we now find that the Sun is going into a period of minimum sunspot activity, and we may be threatened, not with Global Warming, but a new Little Ice Age such as the Earth experienced from the fourteenth to the eighteenth centuries. However, although the scientific community has turned away from Global Warming, politicians continue to act as it's the most important problem facing us. That too has taken on a life of its own.

Example: Environmental regulations require pollution controls that make a percentage reduction in the sulfur content of gasses from coal-fired furnaces. The result is that western coal, which has low sulfur, goes into the highly expensive stack gas scrubbers with lower sulfur content than is in the gasses from eastern coal after it comes out of its scrubbers. Both scrubbing operations cost the same—are very expensive—but while scrubbing eastern coal accomplishes something, scrubbing western coal accomplished almost nothing. The result does little good for the environment, but a great deal of harm to our productivity.

I could continue; but the trend looks clear. As Tocqueville observed a long time ago, the characteristic disease of democracy is not naked tyranny, but a network of fine regulation. Cicero said much the same thing. The end product is not freedom under law, but regulation upon regulation, so that the world might look like India: democratic in form, but with few individual rights; and because elections always go to the incumbents (and election reforms must be voted in by incumbents who are unlikely to reduce incumbent advantages) changes in the trends seem impossible. The ideal of the rule of law fades like dream stuff.

The second disturbing trend is that politics becomes more important than anything else. On reflection, it is not at all clear that who has been elected to public office should affect the life of the citizen more than who has been chosen to head the company he works for, or who has been chosen to head the church he belongs to, or even who has become the head of the local university; but in fact we all know that it's true, that political power "trumps" all others.

Worse, there is the tendency to use political power to destroy one's enemies. The institution of "Special Prosecutor" is now standard, and used as a means to attack anyone unfortunate enough to fall into political disfavor. Thus, politics has become all important; yet it becomes dangerous for any but professional politicians to participate, and if the trend continues, the dangers of being on the losing side become greater.

Scenarios suggest themselves; politicians who are no longer content to lose an election, because they believe that they will be allowed to keep nothing: they will be ruined defending themselves from special prosecutors, and they may be cast into prisons like common felons. In those circumstances the temptation to use political power to subvert the electoral process and thus retain office may be well nigh overpowering. After all, if the Congress can have 98 percent re-election....

The political trends are disturbing, especially to a student of history; but there is one more frightening still.

In 1982 a Presidential Commission on Education headed by Nobel Prize winner Glenn T. Seaborg concluded that "If a foreign nation had imposed this system of education on the United States, we would rightly consider it an act of war." There have been significant changes in the education system since that report; alas, they are changes for the worse.

In 1991, the United States had a lower literacy rate than Iraq. Colleges teach elementary algebra instead of physics. High schools struggle with illiterate students who know no arithmetic.

In his introduction to a new edition of the classic *Teacher In America,* Jacques Barzun says:

> The once proud and efficient public-school system of the United States—especially its unique free high school for all—has turned into a wasteland where violence and vice share the time with ignorance and idleness, besides serving as a battleground for vested interests, social, political, and economic. The new product of that debased system, the functional illiterate, is numbered in millions, while various forms of deceit have become accepted as inevitable—"social promotion" or passing incompetents to the next grade to save face; "graduating" from "high school" with eighth grade reading ability; equivalence of credits, or photography as good as physics; "certificates of achievement" for those who fail the "minimum competency" test; and more lately, "bilingual education," by which the rudiments are supposedly taught in over ninety languages other than English. The old plan and purpose of teaching the young what they truly need to know survives only in the private sector, itself hard-pressed and shrinking in size.[1]

Indeed, we can wish that in 1991 things were no worse than Barzun describes; but alas, we now have high school graduates who not only cannot read at the eighth grade level, but cannot read at all.

The fact is that the public school system of the United States has become little more than a vast welfare scheme for credentialed incompetents; and whatever else the system's output, it will not fail to produce a sufficiency of clients for the "social science" graduates who staff the more traditional welfare systems. Moreover, the trend is down, not up; and every attempt at educational reform runs afoul of the political trends described earlier.

This has serious consequences. To begin with, where will we get the workers who can build the society of peace and plenty described in the first part of this scenario? Indeed, we already project shortages at all levels. The American Association for the Advancement of Science foresees a serious shortfall of Ph.D. scientists. Most industries wonder where they will get their future engineers. At the moment, we meet the shortages in part by draining other nations of their best and brightest, and perhaps that will continue; but assuming we produce enough college graduates and Ph.D.s, where will we get the skilled workers for industry?

Already American industries are finding that in addition to the $50,000 or so in capital required to create a new job, they must also spend another $15,000 or more to educate the new worker to levels that we once expected from high school graduates. The result has serious impact on our competitiveness and productivity. Nor is reform likely: Every attempt at "reform" results in even more money being spent on the system. We are invited to reward those who destroyed the system. This is unlikely to work.

Since this is an optimistic scenario, I am now required to show how we can overcome these serious difficulties.

In politics, I will merely plead the good sense of the American people. We have had great shifts in political trends in the past, and it is at least possible that we will have another; that the sentiment to "turn the rascals out" will prevail at one or another election, and the newcomers will obey their instructions to dismantle some of the structure of incumbent advantage.

For the first time since 1938 the United States is no longer required to be the arsenal of democracy. Few Americans can remember a time when the primary mission of the United States was to stand as the last defender of Western civilization; when we could, quite literally, through miscalculation, misadventure, or sheer funk, throw away the last best hope of humankind.

The strain of that period is obvious. In order to hold together the domestic coalition required to fight the Cold War, Americans, of what is loosely called "The Right," were required to concede a great deal to "The Left." Concessions included a vast bureaucracy, an enormous welfare system, expensive social programs, and the general expansion of the function and importance of government.

Now, suddenly, we can stop and take stock; we can ask the fundamental question, "What is America for?" It was once said "the business of America is business." Whether or not that is true, it has at least become possible again.

The optimist, then, need not describe the political course over the next 50 years; he need only postulate that the pernicious trends will be reversed, the American people will once again get in control of their government, and, more important, that government will become the servant of the people rather than the master.

That does not solve the problem of education. It takes more than mere reform to rebuild an education system when the very notion of demanding success has been lost.

In the twentieth century there have been three enormous systems organized as government control monopolies: NASA, the American public school system, and the Soviet system of agriculture. Each has produced a few early successes followed by a dismal record of failure. Each has utterly resisted reform, whether from within or imposed by external government.

It requires a greater optimist than I am to believe that any organization will voluntarily give up a monopoly.

However, I do note one hopeful trend: Over the past few years, large companies have discovered that it is quite profitable to offer employees day care for their children. Providing day care decreases absenteeism, increases productivity, and creates employee loyalty. The number of companies offering day care to children of employees has grown every year for the past several years.

Postulate, then, that the large companies offer education in addition to child care. Assume further that they do what school systems do not do: that they demand results; that they simply will not accept excuses for failure. "We hired you to teach the children to read. If you cannot do that, we will find someone who can, for we simply do not believe that our children are inferior and unable to learn."

When I grew up in rural Tennessee, our teachers were two-year Normal School graduates; Capleville could not afford four-year college graduate teachers. We had two grades to the room, and about 30 children to the grade; and in Capleville Elementary *every* child could read. Even the village idiot— who was 15 and still in 5th grade—could *read,* if not well. Moreover, by 7th grade we read a number of important works. It never occurred to anyone that there was good reason why we shouldn't.

If the large companies start with the attitude that Capleville School Board must have had when I was growing up, we will soon have students who learn grade school subjects in grade school, middle school subjects in middle school, and on up the ladder. The supply of skilled workers for our industries will be assured.

Of course we can't pretend there won't be problems from having so many of our young people grow up as clients to big companies; but examining that situation is beyond the scope of this scenario.

Meanwhile, small computers have become ubiquitous; and with the computers have come networks. Not only can we "commute" by telecommunications, we can also continue education that way. I need not belabor the point: the next century will be marked by an abundance of information. I said in 1977 that by the year 2000 everyone in Western society will be able to get the answer to any question. It seems clear to me that I was wrong only in placing the year so late. We already have electronic networks that provide information for the asking and we haven't begun to tap the resources we are already technologically capable of. If we can teach the children to read, we can certainly provide them with instruction at every other level.

Thus, I have, I hope, shown a plausible path to an optimistic future.

On the other hand, I have left until last what is perhaps the most important difficulty of all: I have said nothing about the wellsprings of morality.

In the past 20 years, the United States has waged unrelenting warfare on any sign of Western Judaeo-Christian religion. The symbols of religion are excluded from public property, and no hint of religious origin of "values" may be given in schools. Worse: most schools teach, or attempt to teach, a kind of neutrality among cultures and value systems. This not only tolerates, but makes respectable, practices that were prohibited by law only a few years ago. The moral basis for all law is questioned, and often found wanting. We are, it seems, to produce a nation of ethical philosophers who will reason their way to civilized behavior—and do that in schools that cannot even teach the children to read.

The result is not encouraging. We see TV clips of young people accused of rape and aggravated assault, and whose only apparent remorse is that they were caught. Indeed, they don't even seem particularly unhappy to be in custody, since they apparently know that little will happen to them. We may not yet have produced a generation of barbarians within the gates, but surely it is not mere paranoia to suspect that we will manage that feat rather soon. We have sown the wind; perhaps we will reap the whirlwind.

I raise the question, but I have no real answer. Alas, I am not at all sure I understand how we get from where we are to a generation that obeys laws, and does not consider it fun to be beastly. I do believe that education will help; and that education provided by the private sector may yet retain some flavor of the moral principles our courts assure us are forbidden by the First Amendment.

One thing is clear: any optimistic projection of the future must assume that the nation—all of Western civilization—will undergo a revival of morality and find new wellsprings of moral behavior. Indeed, one might even say that this is the very definition of an optimistic future.

Note

1. Jacques Barzun, Preface to the Liberty Press Edition of *Teacher in America* (Indianapolis: Liberty Press) 1981.

FIFTEEN

A VISIT TO BELINDIA

Frederik Pohl

This is the story of Benjamin Brown. He was born in the year 1938, died in 1993, and was born again in the year 2044.

That last statement is not quite literally true. Ben Brown wasn't really "born" again in 2044, of course. That just happened to be the year when the cryonicists were finally able to take him out of the freezer, replace the cancer-ridden pancreas that had killed him in the first place (at the same time performing quite a few other nips, tucks, and replacements in order to take care of other problems that would probably have killed him not much later), and wake him up to tell him that his gamble with cryonic suspension had paid off.

Nevertheless, the fact of the matter was that Ben Brown was alive and well in the middle of the twenty-first century, and if he wasn't really "reborn," what he was was certainly good enough for him...at least at first.

The most important cause for rejoicing, of course, was simply that he was alive. Not only that, he wasn't even in pain; he was comfortable and well fed, and when he looked out the window of his hospital room he could see the wonderful world of 2044 spread out before him. Well, not the whole world, of course, but he could at least see the hospital grounds, and they were bright with flowers and fountains, and everything beautifully green.

True, there were some things that puzzled him. There seemed to be a glass roof over the hospital gardens, and he wondered why that was necessary. Then there was the question of hospital orderlies. When, in the old days, Ben thought about the world he hoped to awaken into, he expected that menial work would be performed by smart machines, maybe even robots. These weren't anything like that. They were just ordinary people. You might even say they were a little less than ordinary, since the orderlies were generally not very big or very strong. They didn't even appear to be very well fed—or so it seemed to Ben, when he noticed that whenever he happened not to eat the rolls, or piece of fruit, that came with his hospital meals, one of the attendants was sure to furtively slip it into a pocket. And the television set in his room was not what he had expected either. It had a big screen, and it was certainly in brilliant color, but it was just as flat as the ones he had owned back in 1993. (Whatever had happened to 3-D television?) And, although his hospital set had more channels to explore than he had time to look at, they all appeared to be showing the same old movies and sitcoms. He couldn't find a single news program to give him an idea of what his wonderful new world was really like.

But when he tried to talk to his doctor about it, she just smiled. "You've got a lot to catch up on, Mr. Brown," she told him, "but it could be pretty disorienting if you tried to catch up on it on your own. So we've, well, limited your access to news. For your own good. We have a staff of counselors right here in the hospital who will help you through the reorientation period. And actually," she went on, looking at his chart, "I see that you're about due for your counseling anyway, because it's about time we started to think of getting you out of here."

That was the kind of news that made Ben forget his questions. "You mean I'm completely cured?" he asked, as astonished as he was delighted.

"Oh, you've been cured for some time," the doctor said, still smiling. "You're even just about healed from all the surgery. We did that for your own good, too; there wasn't any reason to put you through the postoperative pain, so we kept you sedated while your body repaired itself. No, Mr. Brown, we're just about through with you. I'll make an appointment for you with one of our counselors, and I'm sure he'll answer all your questions."

The counselor was a young man with a thick brown beard, and the first thing he said wasn't an answer. It was a question. He studied Ben Brown for a moment, and then asked, "Did you ever hear of 'Belindia', Mr. Brown?"

"Is it a country in Africa?" Ben guessed, pretty much at random, but anxious to be obliging to this young man.

"Not exactly—or, yes, there's a lot of Belindia in Africa, in fact more there than anywhere else. But not the way you mean. Belindia is really the whole world now. The word 'Belindia' was what people used to call Brazil, back in your time. They looked on Brazil as two countries that happened to be occupying the same space, you see. There was a Brazil of the small well-to-do class, who lived at the same standard as, say, Belgium; and at the same time there was the Brazil that was made up of the much larger number of poor people, whose standard of living was down around the level of India's. Belgium plus India: Belindia."

Ben frowned. "That's not the way it was supposed to go!" he objected. "I remember very well that we were trying to cut down on the difference between rich and poor even in my time."

"Did it seem that way to you? Oh, yes, I know that there were all sorts of wars on poverty, and plans to help what you seemed to call 'Third World' countries. They must have sounded really nice and—what was the word you people used? 'Caring?' I suppose they were even intended that way. But they never worked, you know. Not even in your time. All through the last part of the twentieth century the rich kept getting richer and the poor kept getting poorer. Not just here. Not just in Brazil, either, but all over the world."

"That's shocking!" Brown burst out.

"It was inevitable," the counselor corrected. "Think about all of human history. How did poor people ever get rich? There were only a few ways: By going out to makes their fortunes in virgin lands, like the American pioneers; or by finding jobs in new industries and growing with them. But that couldn't happen any more. The poor people couldn't go out to colonize new lands, because there weren't any."

"Not even in space? Not even in—" Brown scowled, trying to remember some of the things he had read about in *Time* and *Newsweek* in his former life. "Not even, you know, artificial floating islands on the ocean?"

The counselor looked astonished. "Of course not, Mr. Brown. The rich countries didn't need to spend money on creating new living space by building space habitats or floating islands. They didn't have any population problems. And the poor countries, of course, couldn't possibly build them, because they didn't have the money."

"No new industries, either? But I remember distinctly that science and technology were going great guns in 1993! There were all sorts of wonderful new possibilities coming out of the laboratories every day! Electronics, quantum-effect devices, fusion power, robots, space manufacturing...!"

He stopped, because the counselor's expression had hardened. "Yes," the counselor said stiffly, "those technological possibilities did exist. They still do, for that matter—if we can ever get them out of the laboratory. But if you want to convert these possibilities into something real it takes money, too, Mr. Brown, doesn't it? Money for research and engineering and development. And there wasn't much of that, because you people soaked all the spare cash up, didn't you?"

"What are you talking about?"

The counselor was definitely controlling his anger now. "I mean," he said, "that you squandered your capital, didn't you? Your governments spent all their money on what you called 'pork barrel' projects—weapons you didn't need, support programs that enriched a few people at the cost of everyone else, dams and canals that served no purpose—and sometimes had to be put back to the state of nature in the end. You used up all your money on things that didn't matter, and so there was nothing left for research."

Brown held up his hand. "I know all about the federal deficits," he said gruffly. "Heaven knows, I had to pay the taxes to pay the interest on what the government had borrowed, and, yes, it did mean cutting back on government financing of research."

"Not just research! You let the whole country fall apart, Mr. Brown—didn't maintain the bridges and tunnels, let the sewer systems and the water supplies in the big cities decay—why, by the beginning of this century the biggest new item in the federal budget was trying to keep the infrastructure from collapsing completely. Not even counting the interest on the debt you'd accumulated. So there was even less money for research and development."

Ben Brown was looking as resentful as the counselor. "All right, maybe the government ran out of money. But what about private industry? They were the ones who did the government work that made scientific ideas turn into things that people could build and use!"

"Yes, it was supposed to work that way," the counselor agreed. "But do you remember junk bonds, Mr. Brown? Do you remember leveraged corporate takeovers? All through the 1980s, especially, there was an orgy of takeovers, with one corporation, or one group of financiers, buying up the stock of another. Of course, they had to pay for the stock. And of course they didn't have that kind of capital, or didn't want to risk it if they did, so they did the buying with borrowed money."

"Well, sure they did! What's wrong with that?" Ben demanded.

"Just that the only way they could borrow the money was by paying high interest rates—which meant they were mortgaging the profits of the corporations they took over to pay the interest on the loans. It didn't make any difference whether the money came from banks, or savings and loans, or junk bonds sold directly to the public—the interest still had to be paid. And the only way it could be paid was for the corporations that were taken over to keep on making those profits—no matter what happened."

"That's what companies are in business for, to make profits," Ben said stubbornly.

"Actually," the counselor sighed, "that's only partly true. Corporations aren't supposed to just throw off a profit *now*. They're supposed to *keep* on making profits—which means they're supposed to invest in new plants when they're needed, and to plan for the future. To invest in research and development."

"Sure," Ben agreed. "Are you telling me they didn't do that?"

"How could they, Mr. Brown? When they had to churn out a profit every year, just to pay off those takeover debts: As soon as there was the slightest downturn in the economy—and there had to be a downturn, with all the savings and loan and bank failures, and the federal debt that kept growing—something had to be done to keep the profits coming. And the only thing that was in the power of the managements to change was to 'cut out frills' in their expenses. And the first 'frill' to go was their investment in research for the future...so, naturally, there were still more pressures on them in future years."

"They shouldn't have let that happen!" Ben declared. "They should have planned for the future!"

"Oh, yes," the counselor said sorrowfully, "they certainly should have done that, but they didn't. So we never did get all those wonderful new technologies you were talking about...except in a few areas, anyway."

Ben brightened. "So there's some good news?"

"Of course there is. How do you think we were able to fix your pancreatic cancer? Medicine's a lot better than it was in your time—they kept funding that kind of research right up to the end, because naturally the people who make the decisions about such things knew that their own lives were directly affected. And then," the counselor smiled again, but not happily, "then there was the other kind of research, the kind people could do in their own basements and garages. We surely had a vast outpouring of new kinds of recreational drugs."

Ben gasped. "You mean *dope*?"

The counselor nodded. "Narcotics, Mr. Brown. Things that you could smoke or swallow or inject into yourself to make yourself crazy. The demand kept getting bigger and bigger, and so the drug dealers kept coming up with new products. The fastest-growing industry in the world, at the end of the twentieth century, was narcotics of a thousand different kinds... and then, of course, there was what you did to the climate."

Ben was getting angry himself now. "Come off it," he snapped. "What's wrong with the climate? I've looked out the window, things look pretty good—"

"In this complex, of course. Under the dome."

Ben remembered the puzzling glass roof over the hospital grounds. He said, backing away, "Anyway, the climate's an act of God, isn't it? You can't blame that on my generation."

"Who else is there to blame, Mr. Brown?" the counselor asked, reasonable. "You people knew in the late 1980s that you were destroying the ozone layer with chlorofluorocarbons—"

"You mean those CFC things the scientists talked about. Sure, but we passed laws to stop that."

"You passed laws to stop *some* uses of them in *some* countries," the counselor corrected. "Then you replaced them with other chemicals that also destroyed the ozone layer—just more slowly—but you increased the volume of those every year, so the damage to the ozone layer kept getting worse, even in the countries that pretended to be taking steps. And most of the world kept right on using the old ones... and you took your time about it all, too, didn't you? So by the time anything serious was done it was too little, too late, as you used to say. Because all the time you were stalling and hoping that you wouldn't have to bother, the CFCs were building up in the lower atmosphere. It took 50 or more years for the things to seep slowly up into the stratosphere where they did the damage. Good heavens, man," the counselor said sharply, "we haven't hit the worst of it *yet!* The stuff you were pouring into the atmosphere 50 years ago is *still* destroying ozone now, and we're all feeling the effects. Skin cancers, cataracts, damage to crops—they're all still getting worse. Along with the acid rain. Along with the wastage and destruction of water supplies—the whole Ogalalla aquifer is bone dry now, because you pumped it out to grow the wrong crops in the wrong place, so now the Great American Desert is a desert again. Along with the global warming...."

He stopped himself, breathing hard. "Sorry," he apologized. "I know you didn't mean any of this to happen. But you didn't do anything to stop it, either, did you?" He was silent for a moment. So was Ben Brown, because he couldn't think of anything to say.

Then the counselor managed a smile. "Anyway," he said, "that's the world you people left us. Belindia. A few million people in the world who live pretty well, under our domes—and a couple billion who, well, don't live very well at all."

Ben Brown's mood had suddenly become a lot less cheerful. "Well," he said, "that's all history, I guess. There's nothing I can do about it now...But I'm alive!"

"You certainly are, Mr. Brown," the counselor agreed. "And you're just about ready to be discharged. About the only thing left for us to do is to discuss your bill."

Ben blinked at him. "My bill? But I put three hundred thousand dollars in escrow to cover all the expenses. That was my whole estate!"

"Of course you did, Mr. Brown. And it very nearly covered everything. But costs have gone up, you know. And you required a lot of special treatment...." He looked at his monitor, thinking for a moment. Then he smiled.

"Actually," he said, "your outstanding debt balance is only about eight thousand dollars. I think we can overlook that. So you can leave as soon as you like—before the end of the day, anyway, so you won't run up any additional charges."

"But I'll be broke!" Ben bleated.

"Well, of course," the counselor said patiently. "You'll be in the Indian portion of Belindia, Mr. Brown. There's no help for that...but at least you'll be alive!" he finished.

And he did not add, *If you can call that living.*

SIXTEEN

A UTOPIAN WORLD

Morton A. Kaplan

The world of 2044 turned out to be the utopia that cynics thought was an impossible dream. Medical advances had extended life expectancy to 150 years. Gene therapy had eliminated hereditary diseases. Fetuses were carried to term in artificial wombs, which played tapes of the parent's physical and emotional patterns.

Brain and cognition research could detect failures during post-partum development of the brain circuits responsible for empathy and ethical behavior. These defects could be corrected before the child reached the age of two, with drugs and social responsiveness that helped the child contribute to its own emotional and ethical maturation. Chemicals were available to aid memory and to facilitate learning. Computer games enhanced reasoning. Languages were learned by the age of seven and physics and calculus by age ten. Native languages were mastered through advanced grammar and writing programs.

Factory and housework was done by robots. Humans functioned primarily in creative (information) roles in the running of manufacturing plants and in the accessing of raw materials, many of them from the seas. Fusion power and very efficient sources of solar power made energy cheap and environmentally safe. The primary economic return stemmed from new products. These were produced by individuals or small cooperative groupings using the resources of extremely powerful and intelligent computers.

Information concerning new products was available through cheap computer linkages. Sales bypassed stores and delivery was by automated systems. Many consumers designed their own products and bought time on computerized robotic systems to produce them.

Holographic plays and movies were called up on home screens or seen in lifelike size in local halls. Advanced computer programs permitted viewers to change both details of plot and the appearance of the characters. This could be done either in explicit detail or computers could be ordered to work out the details on the basis of a few general instructions. The characters could even be ordered to interact with the viewers.

The same type of holographic communication was used for business, educational, or family activities. Except for physical touching, and even this could be simulated, small and extended families could have frequent and cheap reunions.

Similar displays were used to choose and to plan vacations. Different cultures were explored by participating in them within the framework of holographic displays, thus enhancing intercultural understanding.

Children engaged in such displays to learn intercultural tolerance. They also played roles in such displays to gain insight into the requirements of justice. By experiencing the rewards that are given in some roles and the deprivations of others, their natural empathetic abilities were expanded. Such role playing portrayed the role of change in life, thus also enhancing humility. However, even more important, knowledge of physiology and neurology would enable humans to produce in children the feelings of dislike or disgust that accompany social or cultural *faux pas* in cultures other than their own. The impact of this on empathy with other cultures and on the understanding that there is more than one appropriate way to be virtuous was enormous.

Furthermore, since such role playing demonstrated how the violation of local standards does damage, children came to understand that unconstrained individual choice damages others indirectly even where it does not do so directly. By role-playing scenarios in which they manifested unconstrained individual choice, they learned that it destroys that minimal stability of the social and cultural environment that permits stable character and sense of self.

By experiencing different cultural conditions, children also saw that some are better and some are worse, even if no one single way of doing things is unequivocally best. Thus, there still was an emphasis upon improving one's own culture and society and even upon borrowing those features of others that could fit in and improve it. By maintaining sets of limits that alone would sustain stable concepts of identity, the scope of individual choice in those areas of life in which it is genuinely valuable was greatly expanded. Children learned philosophical wisdom not in philosophical terms but implicitly through expanded and realistic role playing.

Parents could have their fetuses adapted to life in the oceans or to some other nonstandard environment. Their hair, eye, skin and other characteristics could be altered in the genes.

Large corporations were a thing of the past because their bureaucracies were maladaptive to the rapid change this world permitted. Individual initiative and creation were enhanced.

Government was far different from that in the contemporary world. A few problems such as pollution or trade or postal exchanges that involve the entire planet required world government. This supragovernmental unit had primary responsibility for monitoring and ensuring human and political rights in the smaller units of government.

Other matters were governed on a regional basis. Europe, for instance, had moved to a weak federal government, weak national governments, and stronger local units. Thus, the

Basques, for instance, had a strong degree of autonomy within a Europe that was not dominated by any nationality grouping. The United States had a much weaker national government, regional units, and stronger megalopolises. The local or megalopolitan units of these weaker federal units had the right to opt into independence of national governments or, with permission, to opt from one government to another.

The advances that permitted these marvels functioned also for the so-called "Third World." Cheap fusion power, cheap information systems, and enhanced learning systems recreated worldwide the human capital which, rather than resources, is the foundation of wealth.

There was no limit on the number of conceptions a couple might have, but, in the absence of death of a child, there was a limit of two on the number who could be brought to term on earth. Additional conceptions would be flash frozen and used to populate other planets or space in some future period in which humankind expanded to the stars.

There was one immense gain that surpassed all the other marvels of this utopia. The combination of enhanced empathy, intelligence, and knowledge made clear the extent to which knowledge is dependent upon background, including worldviews, culture, and physical constitution. The bigotries of race, culture, and religion were finally on the wane, serving as the foundation for the emergence of a lasting world peace.

PART FOUR

REGIONAL SCENARIOS

SEVENTEEN

THE VIEW FROM OUTSIDE AMERICA

Christie Davies

The images of 2044 put forward by American scientists, philosophers, social scientists and writers are of necessity American and reflect American achievements, problems and institutions and incorporate American assumptions, values and myopia. Had a panel of Germans or Indians or Scotsmen or Koreans or Brazilians undertaken the task, there would have been other and different sets of implicit cultural biases to be examined. Americans are no more trapped in a cultural bubble than other peoples; it is simply that in these particular essays, we are confronted with an American (rather than some other national) bubble.

During the period 1940-1990 America was economically and militarily the most powerful nation in the world. The economies of Europe, Japan, and Russia were all but destroyed during World War II, whereas the American economy was immensely strengthened both by the surge of rapid growth that finally pulled it out of the depression of the 1930s and by rapid war-driven innovations in science and technology (not all of American origin) such as nuclear power, jet engines, radar, penicillin, and new synthetic materials. The long Cold War that followed World War II also resulted in an American victory. In America's liberal establishment cowards flinched and traitors sneered at what they saw as American imperialism and the dangerous over-extension of American international responsibilities but ultimately it was the Soviet Union, not America, that collapsed and fell apart under the pressure of superior western economic power and military technology. We all owe our freedom to America's willingness to deploy and, if necessary, launch Cruise and Trident missiles. Subsequent challenges to American hegemony from pettier tyrants such as Saddam Hussein have likewise been crushed with ease.

Meanwhile, though, America's peaceable capitalist competitors in Europe and Japan had recovered from their wartime destruction and their economies have grown at rates far greater than that of the United States. They are now almost as wealthy as America, and are capable both of innovating on their own account rather than copying American technology and of buying up American industry rather than being recipients of American investment. America today is in a position similar to that of, say, Britain in 1880; the top nation but one uneasily aware that technical leadership is steadily slipping away to economic rivals whose citizens are better educated and have a stronger work ethic. *Von Weltherrschaft zum Niedergang?*

As a result of the past half century of American economic dominance, America has naturally received and will continue to receive the criticism, envy and enmity that is the lot of the rich and powerful, however benign their rule. During the next fifty years we can expect even more complaints from those who feel that America possesses a share of the world's wealth out of all proportion to its population and that likewise the American economy consumes a disproportionate share of the world's raw materials. Both charges are meaningless but will provide good excuses for the politicians and intellectuals of other countries whose economies are relative or absolute failures compared to America. Rather than admit the deficiencies of their own leadership, challenge self-destructive national traditions and imitate the dynamism of the Republic of Korea, the Republic of China (Taiwan), Hong Kong or Singapore, they will make fierce speeches denouncing the United States, capitalism and even modern technology.

The fallacy underlying all arguments about the distribution of wealth or raw materials between nations (or indeed social classes, ethnic groups or individuals) is that they take a *static* view of the nature of wealth, capital or raw materials. They regard that complex economic entity, wealth, as comparable to a chocolate cake or a bag of gold coins such that if one person has more, others will have less by exactly the same amount. Capital is seen as a mere assortment of material artifacts that embody labor and raw materials such as tractors, blast furnaces or factories. This crude view of capital is central to Marxism and Third-Worldism alike, but capital isn't like that; rather the value of a capital item is dependent on the amount and sophistication of the knowledge embodied in it. By 1945 the physical capital stock of Japan and Germany had largely been destroyed by bombing, yet within 10 years both countries had re-established dynamic modern economies. Science, technology, knowledge, organization and entrepreneurship are what give capitalist value. Once this is realized it is clear that the distinctions that have been made between 'industrial' and 'information' based economies or between modern and post-modern societies are largely meaningless. Naturally, over time the amount of knowledge incorporated in a machine or stored for future use has increased by steady increments but knowledge, know-how, *Vorsprung durch Technik,* innovation and entrepreneurship have been the very basis of capital and capitalism since the days of Watt, Arkwright, Kay, Stephenson, Faraday and Kelvin. It is knowledge rather than the mere labor of the proles (which could just as easily have been channeled into the building of mausoleums ten times the size of the pyramids for such national heroes as George III, Lord North or Sir Isaac Brock) that *creates* capital. Likewise, there is no such thing as a raw material. What use was iron ore to a Bronze Age man? Why did Napoleon III have the plates at his banquets made of aluminum to impress his guests with his imperial standing and opulence? Why is there a clear upper limit to the price of oil or of bauxite? To whom should minerals belong—to the landowners under whose territory rich deposits have been arbitrarily discovered, or to those who discovered that they were there—how to get them out and how to process them into something useful? America has, in many ways, a better claim to other countries' oil than those who live on top of it. Likewise, the resources of Antarctica and the sea-bed will *justly* belong to those who are the first to calculate how to mine or garner them economically and without creating undue pollution. They are *not a form of collective property.*

Ironically, the Americans have been and will be blamed both for their insatiable demand for raw materials and for the low prices these same raw materials command on the world market. Yet if American corporations should deplete the world's 'stock' of raw materials to the detriment of other consumers, the price of raw materials relative to manufactured goods will rise and the terms of trade will move against America and other industrial countries and in favor of the producers of raw materials even in the absence of a producers' cartel. America's preponderant share of the world's wealth has, on the whole, been just and legitimate since it was the industry, skill, and knowledge of Americans that *created* that wealth. By contrast, we may look at the Soviet Empire where a vast land area, massive reserves of raw materials and very high rates of investment in heavy industry did not lead to innovation, wealth creation or a high standard of living for the people, but only to waste, guns and pollution. It is *only* under socialism that capital has been created, in the manner Marx ascribed to capitalism.

In the Soviet Empire peasants were *deliberately* starved to death, workers were underpaid and overworked, and 20 million slaves toiled in the Gulag to produce a 'surplus' that could be used to make or buy machines or factories. Such a system naturally failed and from this we can conclude that the Marxist analysis of capitalism and of underdevelopment is both fallacious and mendacious. Countries that are poor are usually poor for reasons that have nothing to do with America or capitalism. The very poorest countries are poor because of their past or present accidental or deliberate isolation from the capitalist world as, say, in the case of Nepal, Ethiopia, or Burma. Other countries such as China, North Korea, Romania, Cuba, Mongolia, Albania or Tanzania are poor because of their recent history of dogmatic socialism. Many poor countries in Africa, Latin America, South Asia and the Islamic world may well get even poorer by 2044 because their high rates of population growth will wipe out any economic gains that have been made. It is difficult to implicate America or the rest of the industrial world in this negative cycle unless we make some very odd assumptions indeed about causality and moral responsibility.

In the past, a high birthrate did not necessarily lead to a high rate of growth of population since either it was balanced by a high death rate, including very high rates of infant mortality, or else any rapid rise in population led to a population crisis and an equally rapid crash in the size of the population due to famine or an epidemic of disease. People lived lives that were nasty, brutish, mean and short, in an ecologically balanced paradise of excremental farming and of fraternal sympathy for the animals they ineptly hunted for food. Their holistic respect for the natural order was rewarded by disease, pain, suffering, hardship and premature death. In the industrial world in the nineteenth and early twentieth centuries death rates fell, and after an interval, so did birthrates For the first time human beings really enjoy that *dominion* over nature that God had promised them. Many of the techniques for preventing or curing illness and other causes of early death that emerged from the scientific and technical knowledge and achievements of the industrial countries were exportable, notably sanitation that ensured a clean water supply and the safe disposal of sewage, rapid modern transportation that prevented local famines, vaccination and inoculation, the elimination in many countries of the malaria carrying mosquito, and the introduction of antibiotics and other drugs. This has produced *an entirely welcome decline in mortality* in the poorer countries but also a population explosion. Only an *even more rapid* fall in the birth rate than that which occurred in the industrial world several decades ago can prevent a series of old style population crises and collapses between now and 2044.

As several of the above writers have stressed, new hormonal controls over reproduction may well become the most effective method of reducing population growth in the next decade. The technical problems here are almost certainly solvable as can be seen from the success of the antiprogestin pill RU486 which is safe and effective and already being used in France and China. Other simple, cheap and effective methods of contraception and very early abortion are also currently being developed. The most important task that faces us over the next 50 years is, as has rightly been pointed out, to ensure "that every couple in the world has access to the information and the means to carry out effective contraceptive practice." There is no conflict between the *interests* of rich and poor nations in this point, though it may well be that the funding and the techniques should be provided by the governments and private foundation of those wealthier countries, who already enjoy the luxury of a low birth rate, so that even the poorest couples in the world have *easy* access to *free* contraception as soon as possible. It is no good pretending that some kind of crude wealth transfer is a substitute or trade off for birth control or that it is possible to wait until a growth in living standards slowly makes people begin to think about the possibility of family planning.

For 40 years, under the evil influence of the crudely anti-Malthusian Marxist dictator Mao Tse Tung, the Chinese government claimed that they could achieve economic development without a fall in fertility. They deliberately neglected birth control and the result was disastrous for the Chinese people who now number well over a billion and who, except in a few prosperous capitalist enclaves, remain extremely poor. Since China was, for most of this 40-year period, an autarchic economy having little or no contact with America or other advanced capitalist nations, the latter cannot be blamed for China's plight. The Chinese experiment has destroyed forever the lie invented by Lenin and peddled by his followers that the poverty of the Third World is the fault of the rulers of the more advanced capitalist economies who have somehow managed to frustrate their development. The fate of China also shows that the longer a country's rulers delay giving a high priority to birth control, the more drastic and coercive are the population control measures they are ultimately forced to introduce. The only choices today for countries with runaway rates of population growth are either effective voluntary birth control and early abortion today or compulsory sterilization, late abortion and infanticide tomorrow. There are serious moral objections to the use of the latter methods and these objections are shared by ordinary Chinese citizens who are the victims of such policies. By contrast, a policy of making all methods of birth control and early abortion freely and easily available, widely advertised and free of charge to any person who wishes to make use of them, is not only vitally necessary for economic reasons but is morally the right thing to do.

Women throughout the world should, as a matter of right, (a right as fundamental as any other), should be able to have *free* access to the means of restricting their own fertility and neither the medical profession, nor the lawyers, nor the politicians nor the clerics should be permitted to obstruct this. By the word free, in this context, I mean not only that there should be an absence of restrictions but also that it should be provided free of charge and paid for out of local taxation or international aid. It is the only form of systematic income redistribution that I can think of that is both beneficial (it may even be Pareto optimal) and morally justifiable.

Lest I be accused of being anti-family, let me make it quite clear that I am utterly opposed to the attacks on the family made in many Western countries by a gaggle of privileged harpies and harridans and the male castrati who cling to their bloomers. Also I see birth control as being as much concerned with aiding the childless to have children as with restricting the overall size of families. The effect on population growth of removing the hardship of childlessness from a few couples is utterly trivial and does not contradict the other aggregative and utilitarian aims of world-wide birth control. It is anyway striking that the attacks on modern methods of assisting fertility are often made by the same outmoded and authoritarian individuals who try to prohibit contraception. It is *not* the birth-controllers but obscurantist religious leaders and power hungry politicians who have been able "to enter the bed chambers of three-quarters of the world's peoples and intervene in their most intimate acts" and who have imposed on their own peoples "the dictates of religion and social customs." It was on precisely these grounds that the U. S. Supreme Court struck down the laws prohibiting contraception in the state of Connecticut, thereby setting free the citizens of that state from the moral coercion imposed by the local politicians. There is a clear model here for the rest of the world to follow.

Now it is, of course, possible that even with free and universally available contraception and after the removal of the coercive power of politicians, priests, lawyers and doctors, women and their spouses will choose to have more children than their families (and by aggregation their countries) can support, but we have not yet got to the point where the problem can be stated in these terms. However, during the period 1994-2044 it will (already) make sense to include in any aid package granted to a country with a population problem provision for the education of girls and women to the same level as their male counterparts and for the provision of pensions to widows.

The main choice that will have to be made from 1994-2044 in most countries is between birth control and disaster but there are two other factors that may well have a strange and disturbing impact on the population issue. First, the spread of AIDS is likely, in some areas at least, either to decimate the local population (as happened with the exchange of diseases that occurred after Columbus arrived in the Americas) or to lead to the widespread use of condoms which are, of course, contraceptives as well as prophylactics. Clearly the production and sale of these items should be massively subsidized and the laws against pornography should be tightened up by making it an offense to sell or make a film or video showing unsafe sex in which condoms are not used.

Second, it is possible that within the next fifty years there will become available a cheap and effective way of choosing the sex of one's children in advance. Given the low status of women, the strong preference for sons, and the need to provide dowries for daughters in many overpopulated countries, this will lead to a very sharp decline in the number of female children born. Given that it is the number of women in a society that mainly determines the rate of population increase in a society (the number of men is irrelevant as we can see from the high birthrate of many polygynous societies and the low birthrate of polyandrous ones), this will mean that within one, or at most two, generations population growth will cease. The shortage of women will also speed up their emancipation and will eventually result in the achievement of a society with a low birthrate made up roughly equally of boys and girls. At the same time it should be noted that in societies where women do most of the work and have a high bride-price, parents will prefer to have daughters and in a polygynous or in a promiscuous society, this could lead to a dramatic acceleration in the rate of growth of population.

I have written at some length about the questions of population because for most countries other than America it is the most serious problem they will have to face between 1994 and 2044. The industrial countries of East Asia and Western Europe have low or falling birthrates but much higher levels of population density than the United States and this already has an adverse effect on their quality of life. While some of America's urban overcrowding could in theory be relieved by breaking up urban aggregates such as New York or Chicago and scattering them in medium-sized towns across the empty spaces of the Middle West, this is not an option available to the people of Tokyo, Seoul or Amsterdam. Also for countries such as Bangladesh with low national incomes per capita and a vast population or a high rate of population growth, it is this issue rather than access to the luxuries of American high-tech that will be the dominant one over the next fifty years. In any case technical progress in America may, between 1994-2044, *for the first time in history,* prove to be a burden rather than a boon to the other countries of the world.

The spectacular advances in energy provision, the science of materials, computerization, robotics, genetic engineering and pharmaceuticals that has been predicted will take place in America during the next fifty years will require not only a tremendous increase in the number of both pure and applied scientists and mathematicians but also a highly skilled, literate and numerate wor-force. During the period 1994-2044 America's high schools, which are already in some trouble, will prove to be totally unable to meet these new demands. By 2044 an even greater proportion of school dropouts will be illiterate and few of them will have had any training in mathematics at all. Although it is perfectly possible for a reasonably intelligent pupil to master the basics of calculus by the age of sixteen or seventeen, by 2044 very few American college graduates will have even an understanding of elementary algebra. Their knowledge of quantitative thinking will be restricted to that which is needed to feed someone else's data into a computer on a rubbish in-rubbish out basis. Very few students will graduate in mathematics or science and most higher education will consist of shapeless modular degree programs that will allow students to avoid the challenges of analytical thinking altogether. Even those gifted students who would like to take up such challenges will not be allowed to do so until they are in their late twenties and have already lost their most receptive and creative years. The dislike of the intelligent, which has characterized America since the days of de Tocqueville, will finally triumph by 2044. Its most powerful expression during the next fifty years will probably be through an extension and hardening of institutionalized racism throughout America's universities in the form of racial and ethnic quotas and affirmative action. In the past, American science has been hindered by the anti-semitic and anti-East Asian recruitment and hiring policies of some university departments, but you ain't seen nothing yet. Science and mathematics unlike some of the flabbier branches of the humanities and inhumanities (i.e., the 'something-studies') have two great defects in the eyes of the egalitarian ideologues who control American education. First, there is a clear and unfudgeable distinction between those who are good at them and those who are not. Either you can solve a problem or you can't, and an essay explaining why you can't solve it will not gain the kind of high grades that it does elsewhere in the university, where alphas are sometimes bestowed like old fashioned ticker tape at the whim of the instructor. Second, science is, in one sense, value-free in that it cannot be highjacked for ideological purposes, without in the long run being enfeebled. To deconstruct science is to destroy it. Another way of putting it is to say that science has its own built-in values that provide a basis for resisting ideological manipulation. Scientists may, under pressure, betray their values, but they have a somewhat better chance of resisting political pressure than those of their colleagues who deal mainly with texts and meanings. For both these reasons the next fifty years in America will be characterized by a populist assault on science, the like of which has not been seen since the Indiana state legislature at the end of the nineteenth century tried to fix the value of π (pi) at 3.2 by law. Being irrational folk, the politicians didn't like irrational numbers. Creationists of the world unite—you've only your genes to lose.

Attempts to avoid this dismal scenario will inevitably fail due to the opposition of two immensely powerful groups: the education bureaucracy and the youth culture industry. The educational bureaucracy has an effective monopoly control over an enormous industry with guaranteed customers and a guaranteed income. They have cornered a product to an extent which would have made America's late nineteenth century robber barons go emerald with envy. They are as immoveable, obdurate and obsolete as the loyalist mandarins of the last dowager empress of China. Also, they have a powerful ally in the billion dollar youth industry, which fears and opposes any revival of the abstemious values of work, thrift and learning that might undermine the shallow hedonism that sells clothes, records and entertainment to feckless adolescents.

The coming collapse of American scientific education is bad news for America, but a disaster for the rest of the world: for it means that American business corporations and research institutes will, during the period 1994-2044, be forced to raid other countries for their scientists and mathematicians. The initial losers in this planned and systematic brain-drain to America are likely to be countries such as Britain, Canada, Ireland, Germany, Denmark and the Netherlands where English is either the native language or is widely spoken and understood. Only France and Japan are likely to remain immune from American brain-piracy which is a tribute both to the strength of these peoples' national identity and loyalty and their inability to speak or use (as distinct from read) English. *Banzai la France et vive dai Nippon.* The Yankee privateers and *Seeraüberen* will, however, collect the largest share of their booty from Eastern Europe where there is a large reservoir of badly paid but highly gifted mathematicians and scientists who are now free to emigrate. This act of American I.Q. asset stripping will strangle at birth these countries' transition from military socialism to a prosperous free market economy. Sending a few planes to Russia full of tasteless and emetic U.S. army rations that have passed their "consume by" date is not really a very fair recompense. The only countries that are likely to benefit from losing their skilled personnel are Israel and her Arab and Palestinian neighbors. By 2044 there could well be a hundred thousand former Israeli and Palestinian scientists working together harmoniously in California or New York and there will be no one left who can understand or operate the high-tech weapons necessary in modern warfare. In the Middle East the Americans will have made an intellectual desert and they will call it peace.

To make matters worse, the peoples of Western Europe have already caught the American disease. During the next fifty years bureaucracy, egalitarianism and a hedonistic youth culture will steadily undermine the quality of European mathematical, scientific and technical education. British education is already in a state of serious decline and the recent attempts by governments to respond to this have resulted in a drastic degree of over-centralization, which will make matters even worse in the future. When plundered by the Americans, the British will in their turn become plunderers and will squeeze their former imperial possessions for brains as ruthlessly as the Americans have squeezed them. This has, of course, already happened in the medical profession. British doctors emigrate in droves to America to enter America's second most profitable racket (after law) and

Britain's system of socialized medicine is staffed by immigrants from India, Pakistan, Sri Lanka, Egypt, Iraq, Malaysia, Hong Kong and Ireland. During the next fifty years as the bloodstream of British science is enfeebled, both by the attacks of the American vampire and by self-inflicted anaemia, the British will, in turn, seek their own life-saving transfusion of Asian scientists and mathematicians. The individuals concerned will benefit, for life in a Luton laboratory, in Scotland's silicon valley, or in atomic Aldermaston is still marginally preferable to being stuck in Baghdad, Chandigarh, Madras, Dublin or Lahore and Cambridge is a better place to study black holes than Calcutta. However, the effect on their home economies will be devastating for the countries that have lost their mathematicians and scientists and engineers will be quite unable to take part in the scientific transformations so ably predicted by the above contributors. They will have been condemned to endless poverty not by America's businessmen and soldiers who are the heroes of my tale, but by America's anti-capitalist anti-science, parasitic pacifistic liberal-left of politicians, lawyers, bureaucrats and educationalists who, during the next fifty years, will destroy any possibility of America's achieving self-sufficiency in brains and skills. As usual left-wing self-indulgence and moral posturing will be rooted in gross but unacknowledged exploitation. It will be an economic version of the political link between Alger Hiss and the Gulag. We need a new deal not a New Deal.

EIGHTEEN

THE EVOLUTION OF THE WORLD IN THE NEXT FIFTY YEARS

Jan Knappert

INTRODUCTION

No one can predict the future. The only method of forecasting that carries any credibility is following the trends in the evolution of a society and its technology. The mistakes one can make in this guessing game are illustrated by Camille Flammarion who predicted airplanes with steam engines, and Jules Verne who predicted submarines driven by electricity. Both authors were writing c. 1880, before the invention of the explosion engine by Daimler and Benz in 1886, so they could not foresee that it would replace the steam engine in 75 years.

Nevertheless, the members of the PWPA Netherlands have tried to put together a brief (and still incomplete) survey of the most prominent developments which, in the members' opinion, will influence the world's evolution during the next fifty years.

The general tenor is optimistic about some prospects of human development (world politics, health and medicine), whereas other subjects (e.g., nature versus population growth) left little room for optimism.

THE WORLD OF 2044

This essay has been divided into chapters on a wide range of subjects, to which naturally the specialists who were present contributed substantially, in particular Mr.Hilterman (politics), Dr. Bouwer (health) and Professor Sevenster (the environment).

This writer expresses his deep gratitude toward these scholars as well as Drs. Engelhard and Van der Kreek for their cooperation.

WORLD POLITICS

Whereas populations and cities have grown in such a way as to enable us, by mathematical means, to predict their further growth in the future, the future of the political situation in the world does not evolve in such a regular way. The reason is that there are events in every decade that take the experts completely by surprise. A few examples: The outbreak of both world wars; the Japanese attack on Pearl Harbor; the atom bomb; the rise to power of unknown dictators, for example, the Ayatollah Khomeini, Fidel Castro, Pol Pot, Saddam Hussein; the civil wars in China, Ethiopia, or Yugoslavia; Gandhi's murder; or the dissolution of the Soviet Union.

Suppose before the outbreak of World War II on September 1, 1939, someone had predicted what would happen: Hitler would occupy a dozen European nations; France would be smashed; six million Jews would be murdered; the war in Russia would cost 30 million lives; the war in the Far East would cost many more millions; the war would last almost six years and destroy numerous cities, and so forth. Such a truthful fortune-teller would have been summarily dismissed as a liar and a prophet of doom!

However, since the dismantling of the Soviet empire, it is now possible to be much more optimistic than five years ago. The Cold War is over; the only hostile power in the Eastern Bloc is China but that country is not likely to attack any of its neighbors. Naughty Arab leaders like Ghadhafi and Saddam

Hussein have been taught a lesson and are likely to remain peaceful. Many civil wars will fester on, like the ones in the Sudan and Afghanistan, but they will not constitute a hazard for world peace. The United Nations organizations are gaining power and prestige, so that they are now able to intervene even in civil wars such as the one in Yugoslavia. Governments which are secretly manufacturing nuclear devices have received very strong warnings, such as Pakistan and North Korea, so that they are not likely to use their infernal tools.

All these events seem to point in the direction of greater stability, more peace on earth and more security for people. Equally hope-giving is the spread of democracy worldwide. Dictators who even a few years ago would not hear of sharing power, are now permitting free political parties to organize meetings in public, in the expectation of free elections soon. There are, as always, governments that do not keep their promises, such as Algeria and Burma, where democracy has been suspended, while in the Middle East and China there is oppression.

Still, the principle of democracy as the only truly legitimate form of government is gradually percolating through to even the lowest layers of the most remote populations, for example, in Cambodia.

Another good omen is the apparent willingness of nations who in the past have been at war, such as the United States and Mexico, or Germany, Italy and the rest of Europe, to form alliances like the European Community gradually integrating into a new political entity. Similarly, the tyrannical system of the USSR has been replaced by a voluntary alliance between independent democratic nations, the Confederation of Independent States (C.I.S.).

As a result of greater peace expectations there is a surplus of weapons in the world which are sold off at bargain prices. This is a reason for concern: fortunately nuclear devices will not be sold but destroyed, especially those of Iraq. All this will make the world a great deal safer; but not every country shares in this safety. Some governments make their countries unsafe for minorities (Burma, Sudan), or are partly occupied by other countries (Lebanon, Cyprus), keeping up a state of siege; other countries have armies that need cutting down to size (Ghana, Venezuela), others are ruled by trigger-happy dictators (Libya, Iraq), others, again, continue their attacks on their neighbors (Serbia, Azerbeijan) for reasons that are false and amoral. We hope that soon these governments will be called to heel.

CULTURE

Fifty years ago, the word 'culture' referred to the beauteous and graceful aspects of what people made and did. The most vital part of a nations's culture was considered the Arts, that is, the production of beautiful objects, verse and music. Secondly, the ideal behavior of a person had to be restrained, refined, cultivated, erudite: men had to be courteous, women elegant. Even eating together was a refined art.

In our lifetime, the smiles have vanished or become hypocritic grins. Young faces stare hard at the camera, haughty and bored. Painting has become daubing smears of color on any surface, without effort at creating beauty. Poetry is no longer intended to be either beautiful or intelligible. Music has become loose combinations of screeches and scrapings, outboomed by banging. Novels deal mainly with the problems of lust and fornication. Films have become inconsequential scissors-and-paste showings. Dress, once the object of great efforts to create elegance, was destroyed by Levi's jeans in a capitalist imitation of Chinese drabness. Beautiful clothes are rarely seen nowadays. The art of conversation, courtesy, once universally cultivated as the art of social intercourse, has died. Rudeness is normal. Reading, that is, the art of acquiring knowledge and erudition by judiciously choosing one's favorite authors, has been replaced by moronically watching television, usually devoid of culture. Culture magazines may print photographs of revived folk dances, exotic cookery or the exhibits in some home crafts exhibition. Clearly, the media have conspired to destroy our high culture, because they say it is 'elitist', a Marxist term of abuse.

Culture can be restored if there is a popular will to turn around on the edge of the abyss and relinquish the path of self-destruction. And that will happen only if we discover a new ideology, a new morality, an anchor of faith to hold us in the wild waves of life, a strong hand to guide the ark of humanity.

RELIGION AND MORALITY

As we have seen, societies in all countries will become increasingly secular; in the industrialized countries this process has already gone a long way to cause emptiness in the churches. The process is least advanced in Latin America and the Islamic countries of the Middle East and Africa, where religion prospers.

Hand in hand, with the slackening of religious convictions, we see a diminution of moral rules for which sanctions are no longer considered necessary. Thus we see an increase in practices which only 30 years ago were still considered sinful, such as free sex. We can assume only that this trend will continue in the next half century, as there are no signs that the churches are regaining any lost territory in the hearts and minds of the people. Moral rules, once lost, are unlikely to come back into fashion again, except through a revolution. One such revolution in history was the spread of Christianity in the decadent late Roman Empire. The young church succeeded in reviving the moral sentiments of the Roman citizens through the inspiration of the Christian faith. The Empire collapsed but the Church survived and flourished. Forecasters of the future have to reckon with the unpredictable. In the midst of the decadence of our European and

American societies there may arise a new church, a new faith which may cause people to adhere to the old moral rules again. Perhaps this is what we are witnessing in Russia, Ukraine, Estonia, Poland and other countries recently liberated from the yoke of Marxism and its cynical attitude to religion. The old churches have opened their doors again and thousands, perhaps millions, are crowding into the neglected buildings to worship God and celebrate holy matrimony, whereas in the West, marriage seems to be regarded as an obsolete institution for romantics. Most human beings cannot live without religion, since faith is the fountain of morality, and gives a purpose and happiness to life.

SOCIETY

Society will change dramatically, but not in the same direction in all countries. In the advanced countries, the individualization will continue so that people will become gradually lonelier. Already pocket cassette players with earphones permit a people to listen to their favorite programs, shutting out the world around them. If the institution of marriage survives, it will rarely be for life. Most marriages will be temporary, concluded for taxation purposes or other legal reasons, rather than for procreation. Both spouses will have jobs, often in different places: their children, if any, will grow up in daycare centers and boarding schools. All the work will be done by specialists who will spend their days facing computer screens supervising automatic production lines where robots have long since replaced human workers. Technicians will be responsible for entire factory halls and so, they too, will work alone. Scientists likewise will have a laboratory each and so, they too, will work in solitude. Village shops where social life flourished will be replaced by faceless supermarkets where one can shop and pay without speaking a word. The majority of people will be agnostics: society will be secular. Society will be democratic, but voting will be organized by phone-ins, after which computers will add up the votes: deception will not even be possible as all the votes will be counted automatically. Voters will talk to candidates on their TV or televideo screens via see-phone lines. Candidates will speak exclusively on TV, since there will be no more chance to see them in the flesh, for security reasons.

The masses of unskilled people, many of them immigrants and their numerous descendants, will live in the vast suburb-slums outside the cities.

There will be so much poverty in the derelict townships that no amount of government or private relief organizations can ever alleviate it. Since the group of tax-paying workers will become ever smaller as a result of reduced fertility in the specialist classes, and of increased mechanization, the governments will find fewer funds to supply good housing, roads, lighting, sewerage, clean water and daily maintenance for the poor, whose number will balloon as a result of mass immigration and unchecked reproduction. The proportion of productive versus nonproductive persons will deteriorate over the years and only a steady expansion of mechanization will raise enough capital to keep up with the state's expenses for the poor millions.

In the country, there will be no uncultivated land left: all the land will be exploited, for mining and quarrying, for forestry and for controlled agriculture, which will continue using pesticides and other chemicals to keep production levels.

Class societies, as such, will disappear, but there will still be a sharp division between the well paid specialist and other skilled workers, and the millions of unemployed whose lack of skills will make them unemployable, since there will be no more unskilled labor. All 'manual' labor will be done by machine. Even catering, a service industry which has traditionally employed unskilled labor, will be replaced by automatons producing coffee, soup and hamburgers. Most meals will be bought in frozen form and thawed in a radiation oven, in disposable dishes. Most people will take their meals alone, since their work-hours will not coincide with anybody else's. Cleaning of houses and offices also will be done by machines, including robots and daleks, which can serve tea and food as well. Gardens, too, will be weeded and mowed by robots that can be hired by the hour, making human labor unnecessary.

LAW AND ORDER

In the future, law and order will be ever harder to maintain. Already the number of police officers per hundred persons in the cities of Europe is half of what it was before the war. In addition, the rules of morality have slackened, so that the broad layers of the populace no longer support the police in its efforts to combat crime. Moreover, children are not brought up with the strict warnings that the parents of previous generations repeated to their children: "stay out of trouble, do not mix with those people," and so forth. A very large number of elements from other countries have migrated to the cities of the world. These elements are not criminal as such, but the diversity of their moral ideas puts into question the morality upon which our societies were founded. The Biblical command 'Thou shalt not steal' is put into question as soon as the Bible itself ceases to be the fountain of a society's moral behavior and faith.

As soon as the belief that matrimony is holy and ordained in Heaven began to weaken, so did marriage itself, and with it melted the firm foundation of the family as a pillar of society. The result is, inevitably, that many children are 'let loose' on society with no faith in any authority and no moral discipline.

Another problem of increasing magnitude will be the ever rising density of the population and the vastness of the cities. Police will find it difficult to chase criminals through endless mazes of bidonvilles where the latter are at home. Already there are areas in some modern cities where the police have no power at all.

Conversely, modern techniques such as radio-command systems make it possible for ruthless governments to rule entire countries by terror. There is, however, reason for hope. More countries are now democratically governed than ever before in history, and the first requirement for combatting crime is the people's consent.

FAMILY LIFE

People who married in the 1950s and 1960s were looking forward to a *normal* family life, that is, they were planning to bring up their children at home: only a minority of marriages broke down, according to statistics.

Today, half the children in Britain and the Netherlands are born from unmarried mothers and may never see their fathers. Naturally, the sociologist who is asked to predict the future, seeing the rising line indicating the number of one-parent families, can only conclude that the 'old-fashioned' family will be a minority phenomenon. Today, unmarried mothers are found in families that would have considered it a disgrace a generation ago. In addition, the majority of marriages today end in divorce. In other words, divorce has become normal, resulting in yet more one-parent families. Again, the sociologist has to draw the line further, as the statistics suggest.

Unless a miracle happens, marriage will soon be an old-fashioned minority phenomenon, like a housewife or a big family. As a result of growing up alone and without fathers, children will no longer accept authority or even morality. The mothers, like the fathers, will be absent, working, often in another area: many men are now working in foreign countries and this trend is increasing. Thousands, from Morocco to the Philippines, leave their homes to go and work abroad, so their families do not see them much. The wives have to cope at home. The Asian, African and Latin American peoples will continue to have numerous children, but the Europeans and North Americans will be down to one child per woman. Naturally, this decrease will be compensated for by massive immigration from the three above-mentioned continents, to do the jobs previously done by 'natives'. They will bring their family life with them, as well as other customs and habits, thus completely changing the ethnography of Europe and North America.

WOMEN

The role of women in Western societies has changed dramatically in the past twenty years and it is probable that it will continue its evolution in the same direction. Women will claim and receive functions of power and responsibility which in the past they never occupied. Female prime ministers have ruled Britain, France, Israel, Romania, India and Sri Lanka for the first time in history; Pakistan, it seems, is not yet ripe for this leap into modernity, nor is the Roman Catholic Church ready for women priests; the Anglican Church may soon admit them. Unfortunately, women in business and other stressful positions have begun to suffer from the same stress-related diseases as men: rising blood pressure, heart conditions, migraines and depression. Other concomitant changes in women's lives are that they now sometimes dress like men, and that they no longer have as much time for their appearance. In the future, it is probable that women will become as aggressive as men are reputed to be. Already there is no more distinction in driving habits: both sexes drive equally aggressively. Soon, since no group will stop its progress in one direction until forced to, women might dominate the Western world completely. Men could become householders or be jobless, while the women make all the decisions and execute them. As a result, women will have hardly any time left for procreation and childrearing, so that marriages, if concluded at all, will be increasingly childless. The thus-dwindling population will be replenished from Asia and Africa, and in the U.S., from Mexico, etcetera. As a result, there will be a total change in the world's population. It will become darker and world culture will no longer be 'Northern'. It will become Southern: a new type of person will arise, more emotional, less rational, less business-like. With that, the female population of the world will swing back to the feminine type, tender, peaceful and motherly. The Western world as we know it, will disappear. Faith will rule reason again.

CHILDREN

In some European countries, such as Finland, the birthrate is so low that the government is considering importing immigrants of 'Southern' origin, who are more prone to procreate. The luxury-minded Finns, used to impeccable hygiene, may not be aware that human fertility is bought at the price of poverty which results in lack of child care, hygiene, good education and morality.

While the 'Western' world will have a steadily decreasing number of children per woman, the 'Southern' and 'Eastern' worlds will double their numbers every twenty years by birthrate alone. This surplus of children (five per woman *average*), will inevitably 'spill over' from the 'Third World' To Europe and North America; from Bangladesh to Britain; from Algeria to France; from Turkey to Germany; from Mexico the U.S.; etcetera. The 'boat people' are just one of the already several types of 'refugees', that is, random immigrants, people who just want to be somewhere else. The streams of these people trying to escape from poverty at home will thicken. The Vietnamese boat people are not refugees from an insufferable regime, but from an intolerably high birthrate, which causes poverty and malnutrition. The media distort reality often.

All these migrants (they are not immigrants since they have no legal papers) take their children with them, exposing them to the uncounted hardships of traveling on foot or in an unseaworthy craft for weeks on end without sufficient food or fresh water. Many die. Those who survive grow up in the conviction that only one thing matters: to survive, by whatever means, including theft or deceit.

In Brazil there reigns what is called, incredibly, a children's plague. Born from unmarried mothers and nameless fathers, thousands of boys and girls roam the streets night and day, smoking drugs, selling them, stealing and multiplying, from their teens on. There is no solution for this problem, which is growing in many countries of the Third World. No orphanages can be built for millions of children who are used to a life of freedom.

HEALTH

Dr. S. Bouwer writes that in the future there will be fewer people in wheelchairs, because it will be possible to put them back on their feet again as a result of improved rehabilitation methods. However, the 'developing' countries will not immediately share in this improvement nor in the availability of artificial limbs, or even wheelchairs.

Furthermore, medical operations will be evermore advanced and diversified, so that a higher percentage of people will be returned to reasonably good health. Nor will these operations necessarily be very expensive, provided the governments allow for increased health care and medical attention in their budgets.

Dr. Bouwer advocates the 'return to work' of the ever-increasing number of over-sixty-fives who will be in good health, full of energy, and experienced enough to perform difficult tasks.

Gynecology, birth and baby care will make great progress, so that giving birth will become simpler, and more babies will survive it. Many diseases which are the result of defective genes will be removed through timely detection and early intervention. Although certain operations will be more technical and therefore, costlier, they will also save time which will make them cheaper again.

A problem will be that an increasing number of people will be only theoretically alive, that is, that only their brains will work. This will create ethical problems which the legislators must solve.

Dr. Bouwer expects that many new medicaments will be discovered, even against AIDS and other hitherto incurable diseases. This will relieve the fear of many people that the thousands of immigrants from Third World countries will (it is believed) introduce infectious diseases. Still, health conditions in Europe's growing slums remain a source of concern. Equally disquieting will be the rising number of nervous patients as a result of the strains and stresses in an overcrowded, noisy and complex society.

THE ENVIRONMENT

The term 'environment' has replaced 'nature' as the denomination for the 'wild' world around us, or the land that has not yet become city. It is misleading for it implies that 'nature' is outside us, a mere background, a decor for our acting, whereas we cannot live without nature since we are nature. This nature is being spoiled and Dr. Bouwer expressed concern for the increasingly polluted air that we, and especially our children, have to breathe.

Dr. Sevenster enumerated the many dangers that will result from a continuing rise in the population density in all countries: The ever-expanding cities will cover meadows, marshes and woods under a vast carpet of asphalt and concrete, thus exterminating uncounted species of plants and animals which are vital for the survival (e.g., through fertilization) of fruit trees and other species. Much attention is given to the extinction of the well-known animals, the 'big game and big cats', but the smaller animals, insects and others, as well as all types of plants and trees are of equal, if not greater interest, for our own survival, since we are all dependent on the presence of an unknown number of species.

An equally disquieting result of the rising population is the falling level of fresh water in many countries, from Turkey to the Netherlands. To make matters worse, the remaining water is often brackish, or polluted with oil and chemicals or even refuse. The increase of population and the expansion of human habitation cannot be arrested. Population centers like London, Paris, Istanbul, Mexico, Cairo and many others will continue growing, with or without planning. The rising number of unofficial immigrants will cause vast areas of 'bidonvilles', caravan-camps and slum dwellings, occupying land that can never revert to nature as it will be polluted. These 'immigrant' populations can never be removed, for political reasons, and they will destroy the surrounding woods in search of firewood, since they have no money for other fuel. Over the next fifty years almost all cities will increase their size five times.

TRAVEL

When I was a child, most people walked to their destination, even if that was ten miles away. When I was a student most of us travelled on bicycle and for long distance by steam train. Today, everybody takes the car, even to see a friend on the next street. We are so enslaved by the car that we should be like invalids if it were no longer there. Yet it may soon become too expensive to run and to repair, as conditions in 'Third World' countries already show. It is also possible however, that we will become richer instead of poorer, so that we can afford mini-helicopters, leaving the cars stuck in their endless jams down below in the crowded streets. There will be as many collisions of helicopters as there are now car collisions, but that will certainly not stop people taking to the skies. Helicopters of many different types will be produced in great quantities, probably in Japan first, later elsewhere too. Mass production will make them cheaper than a Rolls Royce.

They will be parked on the vast flat roofs of the countless cities' skyscrapers. Other types, equipped with floaters will bob on the reservoirs which every city will have to dig and maintain for fresh water.

Likewise, for long distance travel, aircraft will have to use lakes for landing and takeoff, since all big city airports are overcrowded and building new runways will take time, cost money by the billions and occupy land near the cities at high prices. I do not foresee floating cities in the oceans, but I do foresee underground networks of shopping malls linked to train tunnels for electric services. Cars are unsuitable for underground traffic because of the fumes. Electric cars will have been developed widely for the purpose so there will be frequent recharging stations in the driving tunnels. At sea the hovercraft and hydrofoil-type vessels will be developed for all regular crossings. Tunnels will connect Britain, Japan, Taiwan, and so forth, to the mainland.

COMMUNICATION AND THE MEDIA

As the number of satellites around the earth increases, so the available news and music will be multiplied for our already saturated data collections and discotheques.

The number of television sets, which after World War II went from naught to one per thousand families, then to one per member of every family, will be diversified: we shall each have a TV screen, a video screen, a video-phone through which we can see whomever we are telephoning, a word processor, a personal computer, and several screens for personal 'hot', lines to communicate with our bosses and colleagues. As a result, going to meetings or attending conferences will be unnecessary, since we can see and hear all the speakers on our own screens. This will make life for businessmen, scientists, engineers, and others infinitely more peaceful, but also lonelier. A personal line and screen will connect a man's office with home, or with his wife's office so that they can communicate at least electronically. Offices may have a bed in them since living space will be increasingly at a premium, and information will flow into the fax systems day and night, for example, share price changes which require immediate attention. Libraries are already becoming video-libraries. The printed word will go out of fashion as a source for information. Home computers will be plugged into main frame databanks so that whoever wishes to know a little more from the business computer, can use the city or university encyclopedic center automatically. Obviously, this information is never better than the brain of the person who fed it in. The advantage of a library is the immediate access to different points of view on one topic. This problem already affects modern news provision: the news reader selects the news (or has it preselected), so that one learns only the news that the national news service cares to reveal. Again, the existence of newspapers with different news and views remains the indispensable instrument of democracy.

LIVING

Fifty years from now, when my grandchildren are in their fifties, accommodations will be many times more expensive than today, so that people (except the very rich) will have to live in much smaller houses or flats with tiny rooms, often sublet to subtenants in order to pay off the huge rents. In the fifty years between the 1930s and the 1980s the house prices in London increased a hundred times, from £300 to over £40,000 for the same ordinary semi-detached house. Prices of office space rose by an even steeper factor, so that modern people will do all their work at home, keeping contact with their colleagues by telephone and video-phone. All the work, including the management of factories will be done by computer. Even now, there are companies in London without offices, composed of three employees (or even less), who meet weekly over lunch to do business. It is already possible to have constant contact with co-workers by means of video-screens so that no one has to meet in person. All this equipment will soon become widely available.

Wives and husbands may even work in different parts of the city or in different cities, so that they seldom meet. This is already common among married consultants, lecturers etcetera. Yet, the majority of people with regular jobs (billions will be jobless) will seldom want to venture out of the city complex where they have their 'studio' (one-room flat doubling as office) because of the insecure streets, the unsafe slum areas around the cities stretching as far as the eye can see. Cities will no longer be surrounded by nature. Life will be lonely. In the city complexes, children will grow up in day care centers because their parents are busy doing their specialized jobs. In the vast slum areas, children will grow up in squalor, lacking everything except company. There will be insufficient education for them.

THE CITIES

During the past half century, both the number and the size of cities have increased dramatically. Large areas of woods and farmland have been swallowed up by buildings, or are covered under asphalt as motorways and parking spaces.

In a continent like Africa, almost all the cities in the interior were built during the last hundred years, for example, Johannesburg (1895), Kinshasa (1898), Nairobi (1901).

Whereas at that time the majority of humanity lived in the country, today this is changing. Already 96% of the British are urban dwellers, and 91% of the Israelis, showing a clear trend.

This trend is in turn linked to the state of development of the country concerned: 65% of South Korea's population already lives in the cities, but in the Philippines only 39%, illustrating that the industrialization of the world will inescapably cause the cities to grow.

During the past hundred years the world's population grew by 100%, from 2.5 billion to 5 billion persons. The big cities however, grew much faster: and the bigger the city, the more rapidly it grew; the result is that there can be no question of city planning. No sooner have the planners designed new quarters for 100,000 people, than half a million have arrived and are living in slums. A corollary of this evolution will be that the pollution of air and water in those cities will increase and never improve, since the measures taken to alleviate this, like improved water supplies or cleaning devices for exhausts will always run behind the facts and figures. An added factor is that the largest cities are also the poorest (e.g., Mexico City with 20 million, Cairo with 15 million, Shanghai with 13 million, Beijing with 11 million and Calcutta with 10 million), so that there will be no funds for any cleaning up or fresh water supplies.

City governments, which in the recent past have endeavored to introduce alleviating social services, have found themselves heavily in debt, like New York City which has 18 million inhabitants. It appears that not only are the biggest cities the poorest, but they are also growing most rapidly, so that they are becoming ever poorer. An added problem will be the huge rubbish dumps, some of which are so enormous that hundreds of people live in them, as one can see in Mexico City and in Cairo. Cairo also suffers from inadequate water supplies, as does Istanbul. Furthermore, the supply of electricity to Cairo is insufficient. In Ibadan there is water only part of the time, and in Khartoum the odors in the air make it almost unbreatheable.

These conditions cause endemic diseases such as dysentery, resulting in infant mortality which will in turn cause parents to want larger families. 'Professional' (i.e., learned) begging and petty pilfering are already normal in Middle Eastern cities. This results in gang formation and eventually (as also in New World cities) in gang warfare to demarcate crime territories. In some parts of these cities the police have no power at all.

As a result, in other parts, where the not-so-poor live, they will organize strongly armed vigilantes to police their houses.

This division between the poor and the affluent has existed since Babylonian times. Recent excavations in India revealed a city of the early 2nd millennium BC where the rich lived in a citadel surrounded by large areas where the poor built their reed huts. In the future, the social divisions will be similar. The cities will contain vast luxury centers where indoor shopping malls provide the citizens with everything they need. They will also have their sports centers and swimming pools in these complexes of skyscrapers merged together by numerous connecting corridors. The entrances to these buildings of the rich will have to be guarded night and day, like ancient cities, to protect them against criminals.

We are not well-informed concerning the size of the cities in Antiquity. A thousand years ago Constantinople had a million inhabitants, so it was by far the largest city of Europe. Traders and craftsmen were attracted to it in great numbers. These commercial and artisanal interests also enlarged the cities of Western Europe in the later Middle Ages, like Paris, Ghent and Bruges.

The first really big city was London in the early nineteenth century when the industrial revolution was in full swing. Industry caused the growth of numerous cities in Europe and the new world. However, most of the cities of the Third World have little industry, yet people come and settle there by the tens of thousands in search of work which is not there. In the future, the cities will be so vast as to cover entire countries, spilling over borders, like the immense conglomerate that stretches from Northern France across Western Belgium into the Netherlands. In the United States where 300 years ago the traveller saw only forest, now one sees only towns between New York and Philadelphia (founded in 1607 and 1682 respectively).

At that rate of growth, easily surpassed by the ballooning cities of Africa and Asia, the future cities will reach from London to Birmingham, from Paris to Amsterdam, from San Francisco to San Diego.

In the Third World, and in some European countries as well, a large part of those agglomerations will be slums, growing so fast that clearing and new building cannot keep up. The largest city in the world in fifty years may well be Cairo with 50 million.

Theft and intestinal diseases will be 'normal': there will be no clean water while the majority of people will be hungry. People already have very little space in the big cities, as is discussed under 'living'. In the future, this living space per person will further diminish. In a city like Naples it is already difficult for pedestrians to walk from one end to the other of a pedestrians-only street. It is as if all traffic slowly solidifies through sheer numbers.

POPULATION

The biggest single problem for the world of tomorrow will be the ballooning population. The number of human beings on earth increased from 2.5 billion in 1950 to over 5 billion today. Even the increase rate is still increasing. It is true that the country with the world's largest population, China, is making determined efforts to reduce the increase, but these efforts are only partly successful. The Chinese government at least realizes that there is a limit to the amount of food available for its people. Egypt is an example of a now common problem: 50 years ago it could still feed its own people, but today, its population having trebled since then, that is no longer possible. Egypt will forever have to live on charity. Food will have to be given to Egypt for free or else that country will face starvation. Japan was in a similar position after the war, but it developed an efficient export industry which enables the Japanese to buy food abroad, as does Singapore.

Saudi Arabia and Bahrain show the highest population increase figures in the world (with Kenya and Burkina Faso), but their oil export enables them to buy food abroad, until that is, the world will no longer need oil for energy.

The situation is particularly worrying in Tropical Africa, where every woman (on average) has seven children, so that the population trebles in every generation. In those countries, a 'generation' lasts only fifteen years, so that in the next half century the population of Tropical Africa will increase *tenfold*. Nigeria will double its population in 18 years to 200 million. The result of this thoughtless reproduction will be increasing famines, which will cause riots, rebellions and civil wars.

The alternative will be massive emigration to the 'rich' countries, which is what happens increasingly in Europe and the U.S.

The number of people who enter the United States territory from Mexico, or by sea, may well exceed a million a year. Only a minority of them go through the legal channels of immigration departments. No one knows their exact numbers, yet they stay. The same is true for the migrants from Northern and Western Africa to Europe, whose numbers far exceed the highest estimates. In France alone, the number of people of (North) African descent approaches ten percent of the population. The majority of these live in camps and bidonvilles in squalid conditions.

In the future, this massive movement of people cannot be stopped. It will continue to grow until governments take drastic measures to return migrants to their native countries. So far, only Hong Kong is consistent in this matter, and naturally so, since it has less space available than any other immigration state.

The alternative will be that the immigrants whose numerical increase is at least twice as high as that of their adopted homeland, will soon outnumber the native European population.

Sociologists have pointed out that in the past immigrants are known to have reduced their birth rates after the first three generations, but this does not apply to the Islamic peoples who form the majority of the immigrants of Western

Europe. They are urged by their leaders to have as many children as God grants. Their women, in total submission, accept their fate passively.

Furthermore, the immigration, legal or otherwise, will not stop, so that unless the frontiers are physically closed, which is impossible, the increase will continue to increase asymptotically.

Most of these people are drawn to the cities, so that there will arise vast areas of bidonvilles around the major cities of Europe and the U.S. Such large slum cities can already be seen around Mexico City, Lima, Cairo and other capitals.

NINETEEN

THE UNITED STATES IN 2044

Gordon L. Anderson

INTRODUCTION

Society in the United States may either deteriorate or improve dramatically by the year 2044, depending on whether cultural and educational trends can be changed in the present decade. There will be plenty of opportunity for people who are well educated and self-motivated, but, in a post-industrial society, employment opportunities for the less educated will decrease. Unfortunately, present trends show the underclass of poorly educated people growing. Further, the breakdown of families and communities among the affluent is psychologically handicapping many of those who should take social leadership roles. Only a change in priorities of the culture as a whole and the quality of education available to the masses will ensure constructive use of our technology and the continuation of democracy.

The future of the United States, like the future of any society, can improve or worsen based on the way the present generation behaves. Throughout most of human history large- scale societies have been dominated by men of military or economic power. The United States was founded on the premise that citizens could throw off tyrants and rule themselves. The experiment has partially succeeded because of the values of Western civilization and self-motivation displayed by Americans in their new found freedom. As the United States has become populated and urbanized it has found itself plagued by many of the social problems its founders and settlers sought to avoid.

The following scenarios, the pessimistic and the optimistic, reflect the extreme ends of a range of possibilities for America. It is likely that the actual year 2044 will be better than the pessimistic view but less utopian than the optimistic view. Americans have traditionally responded late to crises. The educational crisis and the growing underclass in America will probably be met head-on after enough suffering and heartache moves the masses. Although this does not have to be the case, it is in the sloppy style of muddling through, characteristic of democracy.

A PESSIMISTIC SCENARIO

A pessimistic scenario for the United States would follow merely from the continuation of several present trends. It is a society in which democracy is broken down and the powerful few control high technology and rule the weak masses, who have little hope for advancement.

The increased investment in industrial production outside of the United States will lead to greater wealth for the owners of global corporations but fewer jobs in the United States. The saturation of markets will also reduce the number of retail jobs, causing the highest unemployment rates the U.S. has known. The sharp disparity between rich and poor will continue to grow.

The continued growth of the underclass will cause the welfare system to collapse and the resentment and jealousy of the poor will cause uncontrollable violence and destruction to the wealthy and those with jobs.

Gangs will be in control of all major urban areas and will randomly commit atrocities in outlying areas. Members of ruling gangs will not be punished for their crimes. Protection fees will be extracted from residents in their territory.

The wealthy will have to fence and guard their homes. Many of the heads of the global corporations will move to glass bubbles with self-contained ecosystems in the ocean and on polar regions. Such bubbles were proved viable in the experiments underground in New Mexico in the 1990s.

Pollution will be unregulated and increased death related to polluted air and water will occur except for the environmentally pure atmosphere in the life bubbles. Recreation for the masses will mainly consist of virtual reality simulations of rewarding life experiences. Most people, deprived of decent marriages, families and homes, will engage in artificial romance and sexual stimulation with the virtual reality "escape" machines.

The costs of higher education will continue to rise and the number of scholarships continue to decline causing the United States to fall behind Europe and East Asia. The result will be loss of patents and competitiveness. The United States will no longer be considered an advanced nation. Many state universities will be closed and converted to prisons.

High technology centers covering large areas will be run by global corporations. These will have their own space programs, medical centers, and weapons programs. Government sponsored research programs will not exist.

In 2044 the interstate highways will not be passable by automobile. Bereft of funding, they will be unrepaired and overgrown with grass. However they will serve another purpose. They will be cruised by hovercraft, now being experimentally used in Siberia, which will fly forty feet (12 meters) over the ground, water, or snow. These hovercraft will be armed and owned by the global corporations and gang leaders. They will primarily ferry food and supplies. Seventy-five percent of the population will travel on foot or bicycle.

Increased medical costs will force a national health program into existence. The quality of medical care for the poor will initially rise but get dramatically worse after ten years. By the year 2020, over fifty percent of those needing emergency services will die because of delays in getting them service. The wealthy will go around the system and procure medical services privately in global corporation complexes. Many of those with access to private medical services will live 120 years with the help of several artificial body parts.

Democracy will no longer exist as the gangs and the corporate lords will have assumed all government functions by default. Elections will cease to be held in 2020, when the voter turnout is less than three percent of adults. Aspects of government bureaucracy continue to exist but leaders are appointed by those who have secured power through wealth or violence. Procuring government services will be impossible without bribery.

The children of the poor will have little hope of a fulfilled life. Some of the children of the wealthy, reading about the Puritans going to America in the seventeenth century, will form a religious sect which compares the corruption of America to the corruption of seventeenth century Europe. In 2044 they will make a major move to Mars to build a new utopia, free from the vices of Earth. Although they will initially live in a bubble city, the transformation of the Martian atmosphere is already well underway. They believe that their children will live in paradise.

AN OPTIMISTIC SCENARIO

An optimistic scenario sees further equality of opportunity, education, and prosperity for all people. There will be a high amount of leisure time and a shift from acquisition of basic goods to a higher quality of life. "Higher quality of life" refers to lower crime, increased safety of persons and property, a cleaner and healthier environment, increased time for recreation, wide access to good health care, housing and education, and industrial products of increased durability, safety, and beauty. Democracy in a post-industrial age requires highly educated and capable masses to function well.

This optimistic scenario is possible because in the 1990s the United States will realize that it is in a social and educational crisis. Americans, responding typically late to crises, will muster an all-out effort to educate their children and young people adequately for the future. The emphasis on "anything goes" in popular culture will be replaced by the slogan "nothing but the best for all."

The German pattern of universal education and a reduced work week will provide a model for a sustainable high quality society, even though adjustments will be made for the more pluralistic American society. In 1997 the United States will begin an all out program to create universal education from kindergarten through college, with two years of required public service required for all youth between high school and college. Those children at risk because of poverty or broken homes will enter "Head Start" programs which work to offset these negative factors and develop an opportunity for them. Universities will not only teach technical skills but will provide elements of cultural unification by supporting personal and social excellence and opposing ignorance, bigotry, and incivility.

The present school system worked best in the 1930s when the schools were instituted in a relatively homogeneous culture with significant employment in low-skilled industrial and manual labor, and with cohesive family and community structures which complemented the education of the schools. However, the shift to a culturally pluralistic, post-industrial information society, in which the family and community structures are lacking in many areas, requires people with a high degree of knowledge, motivation, and moral integrity. Neither the present pattern of public education, nor the qualifications of teachers, are at a level that can adequately prepare young people for productive life in the twenty-first century. For citizens to be free and responsible voters in a democratic society, they need to be able to confront the great ideas and issues of their age and to obtain good leadership in their society and management of their own lives.

In the year 2044 the official work week will be three days. The saturation of the market in homes, offices, automobiles, computers and appliances has led to decreased demand in the quantity of industrial products. The normal life span of automobiles and of most appliances is twenty years. Because the population of the United States will remain constant after the year 2010, consumption of these items, considered "necessities" by all is only fifteen percent of the rate in the high expansion years from 1950-1970. Food for the nation is supplied by two percent of the population working a three-day week. The clothing industry occupies two percent of the workforce. The housing repair industry will become larger than that of new construction.

The information age will bring the need for a highly literate society which comprehends classical virtues, scientific knowledge, and current social conditions. Although large numbers of computer games will make learning mathematics and other skills more enjoyable, they will also free teachers to focus on the quality of liberal and scientific education. Rote memorization with computer assistance will be practiced individually, while teachers move from student to student, helping each to improve his present level of knowledge. Teachers in large numbers will read the great books of previous cultures and the understanding of human nature and human virtue will move from the elite few to the masses. All students will be reading classics as well as learning higher mathematics and physics in high school. This will begin with the successful schools that the Japanese established initially for their diaspora living in the United States, then become the schools of highest demand.

University education will include life in four different cultural spheres in international university networks, the first of which will be established by the Japanese, in English, to prepare their own young people for life in any part of the world. This system will give the Japanese students an advantage that parents in the United States want to improve upon. U.S. graduates of the first Japanese university network will make this type of college education widely available.

At election time people will be aware of the issues and demand high standards of their political candidates. The 1990s will see the last presidential candidates to attempt to rely on slogans in their campaigns. Any attempt to politicize issues will immediately be sensed by the voters and the candidate will be discredited. Their successors will have to lay out actual plans with timetables to improve society. Those who do not meet the timetables will usually not be reelected.

The political parties will be completely overhauled in after the year 2000 and adopt a platform for the twenty-first century. This will sweep the elections. Politicians will realize that catering to special interests and forming coalitions of opposing groups does not help their campaign. This point will be driven home in the elections of 2000 when more independents are voted into the Congress. Meanwhile, present economic theories will have be adjusted to offset the present oversupply and the growth of an underclass increasingly unable to consume the products.

The platform for the twenty-first century will adopt the principle of subsidiarity which was recommended in Pius XI's *Quadragesimo Anno* in 1931, but largely fell on deaf ears until the large-scale state bureaucracies failed. First was the collapse of communism in 1989. This will be followed by the collapse of the liberal welfare state after the turn of the twenty-first century. With the federal welfare programs bankrupt, the states will be forced to take on increased responsibilities. Politicians will emphasize grassroots democracy which puts welfare on the county and city level. This proves to be more efficient. The role of the federal government will be more appropriate to its level of activity—to provide military security to the nation, regulate interstate trade, adjudicate claims between states, ensure that states meet international environmental standards, and to represent the states collectively at the United Nations.

There will be an initial success of the new politics due to a reduction in federal taxes and thus an increased standard of living for all, especially the middle classes. In the year 2010 the number of employees of the federal government will be one-third less than in 1995. In the year 2025, when federal income tax will be eliminated, the federal government will be reduced to 25 percent of its present size. The decline in federal taxation will be accompanied by a rise in local taxes to cover the costs of welfare. However, because this money was directly supervised by the taxpayers at the local level, it will be spent 300 percent more efficiently. Furthermore, those on welfare will quickly get off, unless they suffer a permanent disability. The net effect will be a reduction of total taxes paid from 57 percent of a middle class income in 1995 to 23 percent thirty years later, the same year that the "Federal Deficit Clock" near Times Square will show a $75,000 surplus for each family instead of the $75,000 deficit each family will face in 1995. In 2025, the federal government will become completely reliant on funds received from the states. The states will be assessed contributions to the federal budget.

Mass education will help the assimilation of minorities into the larger society, reducing racial and ethnic discrimination. Following the revival of the classics and scientific education on a par with the Japanese in schools, community organizations for children will form to reinforce virtuous living. There will be a revitalization of the Boy Scouts, the Girl Scouts, the 4-H clubs, and new programs which reinforce the practice of the virtues in daily life.

The trends toward family break-up in the hedonistic period of the late twentieth century will be reversed as marriage becomes more focused on raising children than on personal sexual satisfaction. Psychologists will be amazed to discover that such families living for a higher purpose actually report higher levels of sexual satisfaction, and happiness in general. American parents, impressed with the support of Asian parents for the education of their children, will reduce their reliance on toys and television as child entertainers and focus on quality experiences for their children.

An interesting result of increased self-reliance on the family level will be the reduction of private debt to financial institutions. Parents will tend to lend money to their children for home mortgages and use the income for retirement benefits. This will reduce surpluses in the financial sector which cause the artificial inflation of the stock market. It will also simultaneously increase the yield to the parents on their investment and reduce the interest rate the children pay on mortgages and eliminate some of the parasitic aspect of banking in the present individualistic sensate culture.

Perhaps no aspect of public life will be more noticeably transformed in the early twenty-first century than the news programs on television. Tabloid style news broadcasts, popular in the 1980s and 1990s will all but disappear from national networks as an educated public demands quality news. The American media will be forced to change as the Japanese and German News "American Editions," beamed by satellite, replace the national network news programs. Only CNN will survive as a major national news network because of its ability to get reporters to sites throughout the nation. C-Span, which will have a monopoly on the activities in Washington, will be watched with great interest and affect the elections of Congress. The educational channels will compile a huge inventory of programs which can be watched on demand through a special viewer dial-in facility.

The local cable television stations, which sat idle most of the time in the 1980s, will become a hub of community life. Highly educated citizens, with more free time, will use advanced video production technology, producing near professional quality programs about projects in the community. Local bulletin boards will help people to network with one another and to cooperate in accomplishing their objectives efficiently.

Ecological consciousness will push the development of cleaner and safer energy. A new generation of nuclear reactors which cannot melt down and do not produce radioactive waste will replace the present aging reactors and the dirty coal plants. The increase of low power electrical appliances will cause the conversion of all open household electrical outlets to safe 12-volt electricity, eliminating the need for voltage converters on appliance cords and the danger of electrocution. The local household current will be supplemented by inexpensive and attractive solar roofing which both covers the roof and provides 65 percent of household electricity.

The interstate highways will contain buried electrical cables from which sensors on electrical automobiles will draw power through magnetic induction. The batteries of the automobiles will simultaneously be charged for use from the highway exit to the local destination. The amount of power use will automatically be billed to the account of the vehicle's owner. Another advantage of this system will be the automatic tracking facility of the automobile that will keep the car centered over the cable and prevent accidents with other cars on the same path.

Airplanes will all be equipped with dual propulsion systems that will allow them to fly into space without being launched by heavy rockets. Trips from New York to Tokyo will take two hours with these planes, which will also be able to make trips to the moon.

One-half of the world's food in 2044 will come from the oceans. Ecological laws which mandate the return to the ocean of one wild fish or animal for each one taken will cause a boom in fish farming in the ocean and also in self-contained systems which recycle the water and can be used to raise fish almost anywhere, including arid inland areas.

The rise of standards of virtue in public life following the education revolution will lead to a dramatic decrease in litigation, legal costs, legal needs, and criminal containment. The number of lawyers will dramatically decrease, as will the cost of medical and auto insurance. The death penalty will once again be eliminated in the year 2015 as a result in the reduction of crime and the relatively few criminals serving life sentences. Greater access to artificial body parts and the hope of an increased lifespan for all will result from lower medical costs and less chance of death due to violent crime or disease.

There will be a dramatic increase in employment in both the fields of education and scientific research. The service sector will be responsible for over 50 percent of jobs. Also the number of artisans and craftsman will increase as people focus on furnishing their homes with original pieces of art and furniture.

NOW IS THE TIME TO CHOOSE

The present trends in the United States show that the pessimistic scenario is rapidly upon us unless we soon shift our educational priorities and cultural goals. The optimistic scenario is completely realizable if such changes are made now. The window of opportunity is in the decade of the 1990s. If we do not choose, our children may not be able to have such a choice.

TWENTY

2044: A VIEW FROM GUATEMALA

Armando De la Torre
Francis Aguirre
Ricardo Bressani
Jorge Arias de Blois
Antionio Gillot
José Asturias
Bernardo Morales
Gerladina Baca
Luis Recinos
Eduardo Suger
Juan Carlos Villagrán

How do we think Guatemala will be by the middle of the twenty-first century?

How would we like Guatemala to be at that time?

How to lead Guatemala so that by that time it would find itself up to our expectations?

A Guatemalan national perspective alone appears to be a very narrow way to catch a glimpse of its possible future, at midcentury after the fifth centennial of the discovery of America.

For that reason, the vision we outline here takes as its minimal horizon the entire Central American region, from Panama to Southern Mexico, although the problems and possible solutions to consider in the areas of nutrition, productivity, education, technology, communications, transportation and the environment are applied particularly to Guatemala, whose peculiarities differ somewhat from the rest of the Central American region.

Although Central America has a population of about thirty million inhabitants, we think it will surpass the one hundred million mark by 2044 at a demographic growth rate of a little below the actual average of 2.9 per year. We also assume that the process of social and economic integration will have advanced enough so that the region will speak with only one voice in the international political arena. Guatemala has today ten million inhabitants and contributes one-third of the regional Gross National Product, and will probably continue to be its center of gravity. Hence, we believe that our tentative prognosis could be applied in a greater or lesser degree to the different political units that, organized as republics, constitute the Central American community.

GENERAL FRAME OF REFERENCE

We can start our analysis from two different hypotheses: either a pessimistic or an optimistic one. As for the "pessimistic," we understand a situation of generalized underdevelopment, which manifests itself in the continuation, or even the amplification, of the breach that separates us from the "developed" world; as for the optimistic, we understand the hypothesis of our being able to surpass that breach, or at least lessen it.

A third possible hypothesis is one that would result from an amalgamation of the optimistic and the pessimistic, but would, in our analysis, be unlikely as well as highly imprecise and in need of a greater rate of specificity that escapes our present concern for a realistic prognosis.

According to the *pessimistic hypothesis,* Guatemala, and the rest of the region at different levels, will be countries very deteriorated environmentally by the year 2044, with high rates of violence and corruption, noticeable cultural backwardness that will not allow the average student to go further than the first three years of primary education, a deplorable housing deficit, high nutritional deficiencies, low rates of savings and investments, and, as a consequence of all this, low per capita productivity.

According to the same hypothesis, the process of urbanization will have involved two-thirds of the total population by then (doubling the actual average), without having been preceded by an accelerated growth in the production of basic foodstuffs that could have served to free up agricultural hand labor and transfer it to the expansion of the country's industrial base. Such a process of urbanization would be just another part of the sad spectacle too familiar to the megalopolis in the Third World: "cities" made over as enormous centers of accumulated marginal population, poor, uncultured, sick and insecure.

All this would not happen at today's pace, but at the pace that by then might have been relatively achieved by the rest of the industrialized world, above all by the gigantic neighbor market of North America.

In this case, Guatemala would continue to be passive receptor of technological transference, with little or no technological creativity of its own; politically dependent on its northern neighbors, and culturally passive and insecure. The country would continue to be in an abject posture before all types of foreign paternalism from every corner of the planet.

In a word, in the pessimistic hypothesis, the leadership that might have risen from within this society in the next fifty years would have failed completely.

In the *optimistic hypothesis,* Guatemala and Central America would, at the very least, have diminished the breach that separates it today from the developed world, at a rate equal to the average of the rest of the Third World countries, which would have taken Guatemala to join the recently industrialized nations of the Pacific Basin.

This optimistic hypothesis presupposes a global market of free trade, in which the comparative advantages of individuals and corporations including their science and technology, will have transcended national boundaries.

In such an optimistic world setting, the permanent trend of prices would be to fall, monetary stability would return worldwide to what it was during the last third of the nineteenth century, schooling would take longer and in the case of Guatemala, we would reach a mean school attendance of nine years (much lower than the world's mean, but much above today's reality), and Guatemala would retain its agrarian profile while achieving internationally acceptable rates of productivity; the rate of our nutrition and health averages would also be very close to the world's mean.

In a word, the Guatemalan situation, and that of the rest of Central America, would be more in accordance with what will have happened over the rest of the planet. For the world at large, we anticipate on average, a better well-being in every sense.

Also, we start with the assumption that history has, by no means, come to its end, in opposition to the renowned theses of Francis Fukuyama, although we agree that history will probably evolve without ideological straitjackets and in a less dramatic fashion, and therefore will be technologically more pragmatic and effective.

We assume too that even though the past is never a guarantee of the future, the concern about the environment will have continued to grow because of the human scale of priorities. It will be by then clearer to everyone that the environmental problem of the industrial externalities, today so heatedly debated within the industrial societies, will have been solved by a new method of exchange of property rights between individuals and nonpublic corporations rather than by increased government regulations.

Likewise, we count on an accelerated massive use of solar energy, that would complement, in an advantageous way, all technologies based on the internal combustion engine and would make feasible the solution of the no less serious problem of the exhaustion of the non-renewable energy resources, all without any decline in the parallel use of nuclear energy.

In energy terms, the standards of production and consumption being registered in Guatemala today conform to the typical pattern of an underdeveloped country. The present consumption per capita of electric energy is low, and the major source of energy in the rural areas—where half the population lives—is wood. But even in this pattern, the technologies that would make for a more efficient use of wood, and its transformation into coal, are now available. For example, dry distillation, pulverization or compaction are on the increase.

We accept as very likely the enlargement of artificial human "habitats," in the oceans as well as in space, from whose technological experience Guatemala could profit in the remote future.

This last fact we believe to be intimately related to the colonization of the Moon, of Mars, and possibly of one of Jupiter's satellites, with the accompanying push of generic technology and the beginning of production of materials in zero gravity.

We anticipate that the permanent revolution in computer science has brought into being a more versatile artificial intelligence and also an excess of information that will force us to design a more careful setting of priorities in our needs for information. This will translate into equally revolutionary changes in the curricula of education at all stages.

We predict that within the coming half century the world demographic explosion (especially in today's Third World countries) will have been rationally controlled, and that the demographic pyramid will have been inverted, all the more meaningful if by then the extension of life expectancy could include a longer, healthier and more useful lifespan that might reach beyond one hundred years of age.

This last possibility would, in turn, mean an exponential growth in molecular biology and related disciplines.

A sounder hope for a longer life expectancy will make education for all a more continuous and interdisciplinary process, with students intermittently coming in and out of classrooms and laboratories, and even through systems of long distance education.

Certainly, we don't know which leaders as persons or as groups, or which regions and countries, will become more dynamic through competition over the coming half-century, but this future looks to us to be at once both promising and challenging. However, it seems that the feasibility of any trend toward a better future will rest more in the hands of those who can contribute fundamental solutions through their mastery over matter and the various forms of energy than in the hands of those who speculate about the "big picture" who serve men through their philosophical reasonings—men, that is, who are always restless and always creative.

All of this will mean a larger depolitization of public life, which will allow for new forms of religiousness—probably more intimate and personal—and for a larger experimentation in the arts as well as a deeper respect for common, not positive law, while the volume and scope of positive legislation is reduced. In other words, the human individual might return for his due, as some exceptional spirits suggested during the Renaissance, and as was dreamed by some men of the eighteenth century Enlightenment. On a human scale, the possibilities of change for the better look, at present, almost infinite, even for a Guatemala and a Central America still struggling with underdevelopment.

GUATEMALA AND THE REGION AS WE WOULD LIKE TO SEE IT IN 2044

We would like to see Guatemala's society integrated into the First World.

Presently, only a small fraction of the population enjoys the rates of caloric intake, health care, decent housing, quality education, and overall liberty under the law comparable to the ones enjoyed by the middle or upper social classes of the developed world.

Success in integrating Guatemala will be grounded necessarily in its agricultural modernization. This means that our country must sell its products in international markets at high value added.

To accomplish that, it becomes essential to concentrate our scientific and technological progress in those areas in which we have more possibilities to compete successfully.

The majority of our countries show enormous potential in the fields of agriculture, stock raising, and forestry. The key lies in raising productivity per capita and per hectare. Through these measures we would not only solve the burdensome problems of food availability but also the problem of access to good nutrition. This will also facilitate the modernization of the secondary and tertiary production sectors.

The starting point for a rise in production would have to be an efficient scientific-technological structure and the achievement of a no less efficient system of transfer of knowledge to the agents of production. In the long run, the size and dynamics of the industrial sector have been in most countries a function of the growth of agricultural productivity, productivity which puts at our disposal enough food surplus for a growing population, plus more abundant raw materials and scaled up industrial and economic ingenuity that favors their transformation into manufactured goods for the satisfaction of an ever-growing consumer market.

The strategy to follow should be to obtain for Guatemala a solid place in the international division of labor, or at least to actively participate in the world market of goods, services and knowledge with the maximum possible flexibility (preferably the flexibility that issues from the spontaneously self-correcting market), and without having recourse to the inefficient system of so-called import substitution schemes.

Certainly, we know that our space in the international economic system will not be given to us free of charge.

Given the tropical conditions of the country, with a high degree of photosynthesis all year round, it is very likely that the technological areas in which we will be more successful are biotechnology and genetic engineering applied to our natural resources and agricultural productivity.

For this, more investment at the base is needed in human capital and the endowment of centers for top-quality scientific and agricultural excellence. Therefore, changes in teacher education should not be restricted to the universities and institutes geared to investigation but widened to include primary and secondary education. Technological awareness should be reinforced among the productive sectors and the leading political groups too.

From another perspective, free market measures will have less effect if we do not pay special attention to education in science and technology. Knowledge obtained through scientific research has become not a luxury but a necessity by 2044. The accrued rate, on a worldwide scale, of the available technology is very accelerated indeed and unfortunately Guatemala today is not yet well equipped to select and/or to adapt that technology at the same rate at which it is produced.

One of our priorities , therefore, should be to increase our research capabilities at the same time we increase our capacity of doing profitable business and building profitable research facilities.

Research should be simultaneously basic and applied. For example, productivity research would include investigations into the usefulness of the medium (soil, salinity, pH, etc.), its resistance to disease; into the efficacy of plague control; into more efficient photosynthesis; into animal development, crops cycles, the use of mutants that can fix nitrogen in the air or that can prosper in salty or acidic soils. All of this would dramatically increase the volume and quality of food production for the basic Guatemalan diet (beans, corn, potatoes).

In the area of product development, research would concentrate on the nutritional quality of food, on the industrial quality of what is manufactured, on commercial quality (lowering cost, not raising prices), on the increase of caloric intake, flavor, specific organic substances, medicines and the diversification of its uses.

Other promising fields are constituted by the refinement of patent systems, the training of a better qualified laborer, faster automatization of production, environmental rehabilitation (for example, well-planned reforestation, and recycling campaigns for solid waste) or the use of substances of local origin in the pharmaceutical and cosmetological branches of industry.

To all that we would have to add a more receptive attitude for technological innovations that come from outside the region, because we are conscious of our limitations in these matters (unlike some people in the protected market of technology and science of Brazil, Mexico and Argentina).

The goals to achieve by the year 2044 would include, among others, from the scientific and technical perspective, the following:

1) to secure a sufficient and stable nutrition model for all the population;

2) to eradicate most forms of malnutrition;

3) to make more effective the transmission of agricultural technology;

4) to lessen the gap in well-being between the various regions of the country

5) to increase the consumption of animal by-products (here, genetic engineering applies not only to agricultural products, but also to traditional animal breeding, which would involve the use of clonic embryos and of "surrogate mothers" through artificial insemination, which would allow, in a shorter term, the introduction of new and desirable genes);
6) to diminish the growth rate of the population;
7) to economize, more efficiently (via the market) on land, water, and energy resources;
8) to improve the quality of housing and health care; and
9) to make investors more conscientious of the profitability of science and technology (research and development).

Beyond genetics, we would like to see in Guatemala an authentically developed modern culture, by which we mean the acculturation and assimilation of all those values that human experience has shown to be more beneficial to humankind.

Such an updated culture would revolve around the individual's intrinsic values and the concept that law exists to protect and not to control the individual.

We believe all this to be feasible to the extent that people succeed in overcoming their atavistic tendency to violent conduct. Guatemala bears the burden of a sad history of violent abuses. The ideal of the rule of law, that is a government of laws, not of men, has only been realized in part. We aspire to witness a genuine revolution in legal philosophy by which we govern ourselves, an effort that probably will be more exhausting and risky to us than the one setting the competitive basis for our incipient genetic technology.

We are countries of high indexes of male parental desertion, the consequences of which are reflected in the poverty of vocabulary of the man-in-the-street, in the blunted intellectual curiosity of our well-to-do, and in the generalized inability to postpone gratifications, which necessarily make us less adult countries, and socially more disorganized.

That education for the rule of law in the twenty-first century will have to start at home and be confirmed at each and all levels of formal education. We shall have to discover the ethical foundations of a competitive society and start preaching by example.

One forgotten aspect in our educational system has been the training of intermediate technicians in the fields of electronics and in the management of capital goods for production, as well as the solid formation of engineers in all other branches of engineering (civil, systems, electric, chemical, industrial, etc.).

We believe that by the year 2044, the countries now classified as underdeveloped will be divided into two divergent groups:

A) the ones that passively have received their technological "quantum" and whose development, therefore, would have been slow, as if by inertia, and;

B) the ones who are aggressively aligned with self-development, which will result in an accelerated well-being for their population.

In both cases, the point of reference will have to be their present state of development, not the development of those countries that have gone ahead of theirs in the race for progress.

At the end of the next fifty years, therefore, there might be countries which still will define themselves as farms and mines for the developed world and others which might have "taken off," incorporating themselves, by applying the cutting edge of science and technology, into the First World.

For the first group, the North-South conflict will continue to loom as a sad reality; for the second one, the world will have been reduced to a glove in which its four cardinal points will be mere geographic references, not an axis of conflict.

It is for this second group of countries where the already mentioned rule of law turns into an ideal that is within the grasp of the Guatemalan.

Its most obvious effect will be a permanent and generalized peace, in which the contribution of the multinational corporations will help pioneer the attainment of a balanced compromise between productive expansion and the concerns of the environment.

We believe that peace which issues from the rule of law will add decisively to the improvement of income per capita measured in real purchasing power, to an enlargement of scholastic achievement over an arithmetic average of nine years, and to an enrichment of culture measured by larger international imitation of models generated in our lands.

Peace, likewise, will allow a better use of classic fuels as well as a more extensive use of plastic materials in combination with traditional materials like cement, lime and steel, without much detriment to the environment.

Peace will also deepen the very important respect for the rights of property (of land, of profits, of inheritance), as well as intellectual property.

Peace under the rule of law, additionally, will allow us a more efficient use of new communication techniques for educational ends, which will facilitate the penetration of the concept of the importance of technology in the great masses of our population, accompanied by a greater intellectual eagerness of our young people (the healthy restlessness to know the how, the why, and the what for of technology), as it is wisely stated in Article 80 of the Constitution of the Republic of Guatemala.

Peace under the rule of law will allow us to make a qualitative jump in higher education, where we might succeed at last to balance a humanist formation of the whole man (through philosophy, history, literature, arts, music, and religion or ethics) with other endeavors more appropriately scientific and technological (through mathematics, physics, chemistry, biology and the applied sciences).

This will probably lead, at the same time, to a larger demand of educational centers that will tend to organize in accordance with their diverse specializations, as appears to be the case today in Japan, and with a more intensive use of electronic media, of research laboratories and field work.

WHAT TO EXPECT FROM OUR LEADERSHIP

By the year 2044, we will have among us leaders with a long-term vision (the ones with special academic devotion, or civic or religious qualities about them), and the ones with short-term vision (usually the ones in party politics and members of pressure groups, and above all labor unions). From the first set of leaders we should have accomplished the setting of goals initially unpopular and revolutionary; from the second set, the practical task of motivation and persuasion of the great masses.

A lot will depend on faith and security in oneself, that will have been reinforced through the value system transmitted at home, school, and the street, and on the love of risk-taking and pioneering adventure, through which society, as a whole, rewards or discourages the most enterprising spirits in its midst. All this also will have contributed to the fact that our society's arithmetic mean will be younger and that we will have gotten accustomed to a permanent dialogue —increasingly more intimate—with those societies that have rushed ahead through their own scientific and technological élan.

Equally useful will have been the assimilation of an ethnic, religious and philosophical pluralism so rich in Guatemala, the best broth for cultivating the emergence of vigorous, original, and inevitable creative personalities. This amounts to elevating mutual tolerance to the supreme place among social virtues.

Last but not least, to have reached a rate of institutionalization in public life such that those men who have been charged with the ultimate responsibility for the decisions in the short and long terms, affect that general well-being show themselves to be readily receptive to the leading ideas in scientific, technological, religious, artistic and philosophical investigations carried on in teaching and service centers.

We Guatemalans who sign for these thoughts about our future, reaffirm our initial optimism paraphrasing a statement—in another context—by George Bernard Shaw:

"I see the world as it is and I ask myself, why?; and I dream about how I know it could be, and I ask myself, why not?"

TWENTY-ONE

THAI SOCIETY IN THE 21st CENTURY: THE QUEST FOR WELL-BALANCED DEVELOPMENT AND PEACE

Weerayudh Wichiarajote

BACKGROUND OF THAILAND

Centrally located in Southeast Asia, and internationally recognized as the regional hub, Thailand is increasingly becoming one of the significant partners in the shaping of the Pacific Rim Era in the forthcoming twenty-first century.

At the end of World War II, Thailand was basically an agrarian economy, heavily dependent on rice which accounted for about 25 percent of the Gross Domestic Products and about one-half of the exports. At that time, Thailand had a very small manufacturing sector and limited infrastructure.

However, over the ensuing 40 years, real annual GDP growth has averaged about 7 percent. Consequently, the infrastructure has been enormously expanded, economic outputs and exports have become rapidly diversified, and poverty has been substantially reduced. During the remarkable economic performance of the last five years (1987-1991), Thailand has recorded the highest annual average growth rate in the world, and is still flourishing well into the 1990s albeit at a somewhat lower rate after the political unrest in 1992.

The unfortunate bloodshed incident resulting from this political upheaval has awakened a wide spectrum of political awareness among the general public, particularly the rapidly rising middle class. Now, there is a widespread public consensus that the political system of Thailand should be reformed in such a way as to allow more participatory democracy, based on a moral foundation, to grow at the grassroots level, and to safeguard against military intervention in political affairs. Consequently, in July 1992, the Anent Panyarachun's Government encouraged the Committee on Political Science and Public Administration of the National Research Council to formulate a national plan for political development to be submitted to the government for approval and further action. If this well-intended plan for political reform is realized, social-cultural and economic development would be effected and accelerated at the local level, thus providing a broader and more stable base of social infrastructure for national development.

Toward the end of the twentiethth century, Thailand has become more and more involved in the rapidly-growing global economy, mainly because of the accelerating effects of industrialization for export purposes. Thailand also has become an international player in the complicated Indochina politics, as well as the fast-changing world political arena, in aspiring for a better and refined form of democracy, which has become the champion of the world ideology after the collapse of Communism in the USSR. By the end of this century, Thailand should be well-prepared for becoming a partner in development for the forthcoming Pacific Rim Era.

Furthermore, in the present decade of the 1990s, Thailand hopes to reach the goal of being one of the Newly Industrialized Countries (NICs), in the same category with the four Asian "Tigers," namely, South Korea, Hong Kong, Taiwan and Singapore. Viewed from this perspective, the "May Political Reform" which may give an impetus to accelerated and integrated spiritual-social and economic development would definitely render a better opportunity for Thailand to become a well-qualified NIC.

TWO ALTERNATIVES FOR THAILAND TO BECOME ONE OF THE NICs

One fundamental question raised by some Thai scholars is whether we really want to become one of the NICs which developed by emulating the economic development model of the industrially advanced nations favoring material development far beyond or at the expense of mind development.

However, one may retort "What's wrong with this model of economic development?"

Weerayudh Wichiarajote's article entitled "An Alternative Approach to World Development" published in the *International Journal on World Peace* (Sept. 1991, vol. VIII, no. 3) could well clarify the dangers of this prevalent global trend of economic development which is viewed by the author as "the Greatest Mistake of the 20th Century:"

> The greatest mistake of the 20th Century is probably the promotion of imbalanced development, with an out-of-proportion emphasis on the material dimension of development, through efficient and effective application of scientific technology.
>
> This trend of imbalanced development has left social and mind dimensions, the other two essential components of life and social development, lagging far behind. Consequently, this misdevelopment has brought along with it many social-psychological ills, such as raw greed, tension, neurosis, moral degradation, leading to a spiralling of selfishness, social problems, crime rates, broken families, social conflicts, confrontations and wars.

These social-psychological ills and conflicts, brought about by imbalanced development with the leading role of natural sciences and technology, have more or less given rise to the bloody May political turmoil. They also have daunted quite a number of Thai social thinkers who often made a quip that "NICs" is abbreviated from "*Narok (Hell) Is Coming Soon.*" It is then imperative that Thailand thoroughly and wisely consider alternative models of NICs, as well as ways and means to *effectively* and *peacefully* realize the chosen goal.

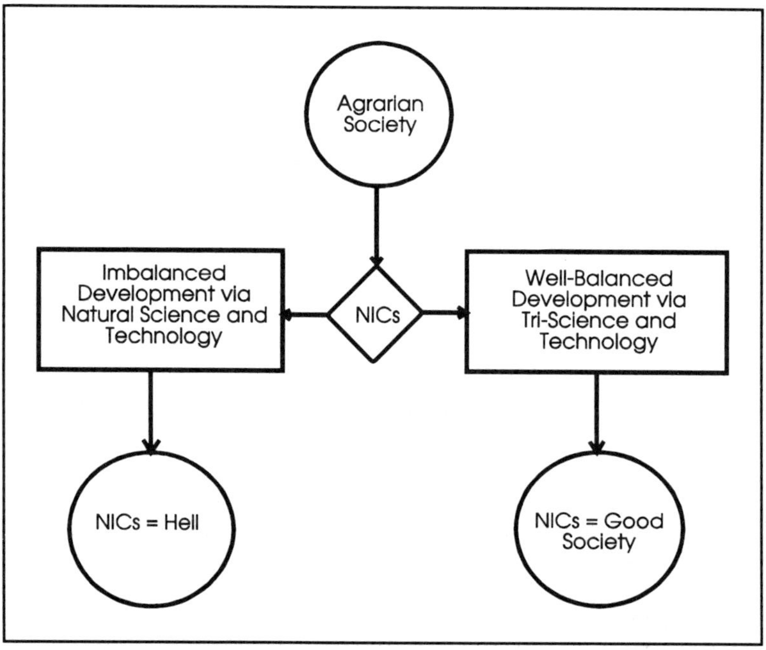

Figure 1 Two Alternatives for Thailand

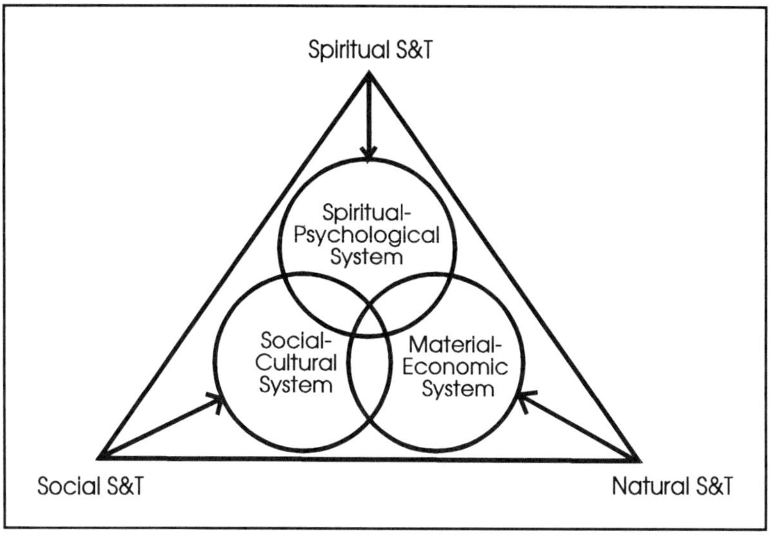

Figure 2 Three Systems in Our World which should have Balanced Development

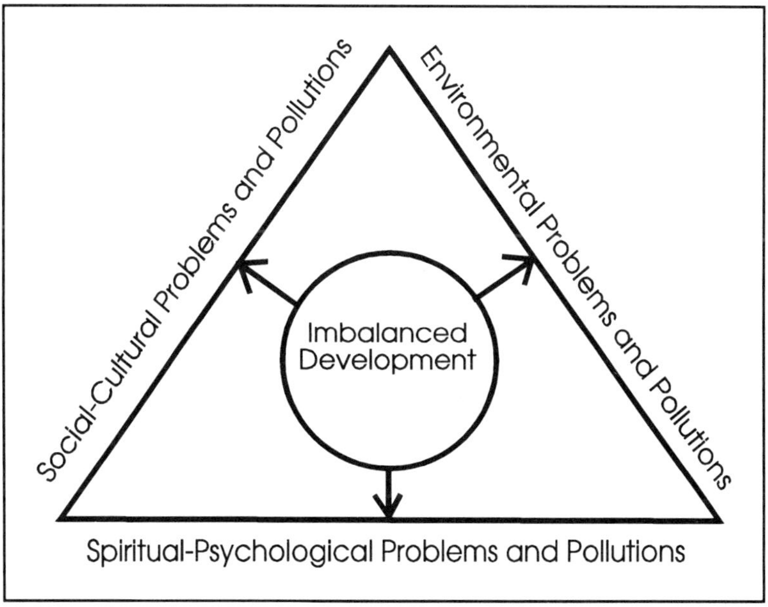

Figure 3 Potential 3-D Dangers to Society

For about three decades, Thailand has adopted this imbalanced development that is widely practiced in the modern world. But recently, and particularly in 1985, Thailand began to realize the dangers of this misdevelopment by gradually reshaping the National Development Plan toward more balanced development. This shift of development policy has already been substantiated by the government in proclaiming a *Well-balanced Development Ideology* known in Thai as "Pandintham-Pandintong Ideology," or "The Land of Virtue and Wealth Ideology" on May 5, 1985.

It is apparent that Thailand has more or less chosen the well-balanced alternative as its approach to national development. Through this approach, Thailand would definitely achieve very unique national goals within the future of 10 to 50 years, provided that the succeeding governments are committed to policies which promote national development in accordance with the Pandintham-Pandintong Ideology.

TOWARD THREE-DIMENSIONAL NATIONAL GOALS

What, then, would be the unique development goals of Thailand in the year 2000, and 50 years from now?

Recognizing that three basic targets for development are Man, Society and Nature/Environment, taking into consideration that three basic dimensions of development are Spiritual-Psychological, Social-Cultural, and Material-Economic, and provided that Thailand is serious in the long run, in promoting national development in accordance with the Pandintham-Pandintong Ideology,

It is therefore feasible for Thailand to envision three-dimensional goals of life and society illustrated in Figure 4.

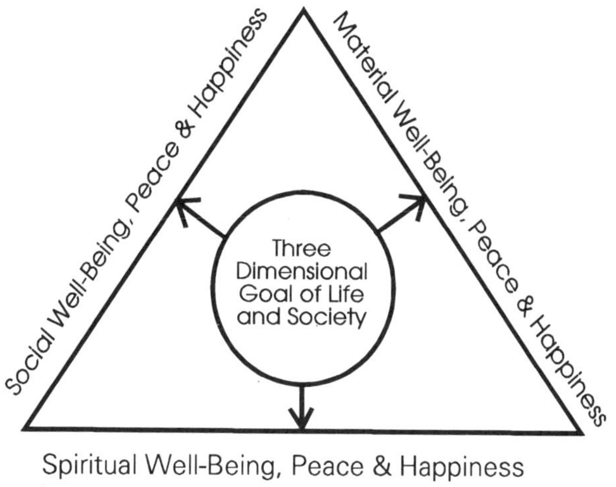

Figure 4 Three Dimensional Goals of Life and Society

TRI-SCIENCES & TECHNOLOGIES FOR THE ACHIEVEMENT OF WELL-BALANCED DEVELOPMENT GOALS

It would be most unlikely that well-balanced development goals could be achieved solely through the application of natural sciences and technologies. On the contrary, they represent the major factor in producing imbalanced modern development based on material emphasis. What then would be an alternative to the emphasis on natural science and technology which would be appropriate for the achievement of well-balanced development goals? The Author would like to propose an integrated and multi-disciplinary model of Tri-Sciences & Technologies for this purpose (see Figure 2 above). "Spiritual Sciences" in this case embody three disciplines:

1. Science of the Human Mind
2. Science of Behavior
3. Science of Morality

According to this definition, Buddhism can be categorized as an integrated spiritual science which basically deals with the three areas of life in a multidisciplinary manner. Psychology should be included mainly under Science of Behavior, and partially under Science of the Human Mind. Science of Morality deals with the law of ethics and the law of moral conduct as discovered by Lord Buddha about 25 centuries ago. As for spiritual technology, this deals with the application of spiritual-scientific principles for the promotion of development and solving the problems of life. For instance, "peace of mind" can be developed through "Meditation Techniques."

Social Sciences and Technology usually embody such disciplines as sociology, anthropology, political science, business administration, and education. The application of social-scientific principles for the promotion of development and solving the social problems of life and society would constitute social technology. For instance, QC Technique is a social technology.

Natural Sciences normally embrace physics, chemistry, biology, zoology and other related sciences of the physical and biological world. The application of natural-scientific principles for the promotion of development and solving the material problems of life and society would constitute material technology

THAILAND'S DEVELOPMENT SCENARIO IN 1994
RESPONSES TO DEVELOPMENT CHALLENGES

The second half of the twentieth century witnessed the rapid social and economic development of developing countries. Thailand met these development challenges by launching a systematic national development plan for the first time in 1961. Since then, Thailand's laissez-faire economy has been transforming into a free-market one. With dynamic growth, urbanization began to proliferate and the cityscape began to change. This was marked in Bangkok and other major regional cities which serve as commercial and industrial centers. More and more cars appeared on the roads, new buildings sprouted up everywhere and Bangkok began to spread in all directions like any metropolis. This modernization process brought in new lifestyles, as well as created consumer spending habits and better living conditions for the newly emerging middle class and the well-to-do. At present, Thailand is in the Seventh Five-Year Plan.

Throughout this period, national planners concentrated on channelling the nation's growth during the First, Second and Third National Development Plans by developing basic infrastructure which formed a solid foundation for the impressive growth of the country. In later Plans, Thailand embarked upon industrialization by first adopting a policy of import substitution and later a policy of global export. Its prime economic development areas were designated as the Eastern Seaboard and the Southern Seaboard Development Projects.

RESPONSES TO GLOBAL MEGA-TRENDS

John Naisbitt and Patricia Aburdene (1991) reported the following new Global Mega-trends in the next ten years as "Gateways to the 21st Century":

1) The Booming Global Economy of the 1990s
2) A Renaissance in the Arts
3) The Emergence of Free-Market Socialism
4) The Global Lifestyles and Cultural Nationalism
5) The Privatization of the Welfare State
6) The Rise of the Pacific Rim
7) The Decade of Women in Leadership
8) The Age of Biology
9) The Religious Revival of the New Millennium
10) The Triumph of the Individual

THE BOOMING GLOBAL ECONOMY OF THE 1990S

In the decade of the 1990s, the world is entering into an unprecedented period of booming global economy in which Thailand is one of the leaders. In recent years, Thailand, now becoming a more and more export-oriented economy, has been able to achieve high growth rates ranging from 7 percent to 13 percent. Thailand has been an active member of ASEAN, or "Association for South-East Asian Nations" comprising Thailand, Malaysia, Singapore, Indonesia, the Philippines, and Brunei. Thailand submitted the well-supported proposal of "ASEAN Free Trade Area" (AFTA) in the recent ASEAN Summit Meeting. Regional economic groupings are now in the formative stage in many parts of the world. In the long run, these Regional economic groups will merge together, forming larger groups, and finally becoming a single but global economy. Definitely, Thailand will fully participate in this emerging Global Economy.

THAILAND'S ROLE IN THE PACIFIC RIM ERA

The saturation of economic development in the Western countries, the collapse of Communism in the former Soviet Union and its Eastern European satellites, and the formation of the North American and European Economic Communities serve as indicators for the reorganization of the world political economy, in general, and the Pacific Rim political economy in particular.

Although Japan has been expected to play a leading role in the Pacific Rim, Thailand is expected to be a regional leader in Southeast Asia with respect to economy and politics. Thailand is strategically positioned in such a way that it can best serve as a regional intermediary to bring about economic and political integration in Indochina as well as the whole region of Southeast Asia.

To accomplish such a role in this decade and in the next century, which has been called the "Pacific Century," Thailand would need to prepare leaders adequately for this coming

Pacific Century by offering services for higher education which would be open to students in particular from the Pacific Rim countries, as well as from the whole world in general.

THE RELIGIOUS REVIVAL OF THE NEW MILLENNIUM

The trend in Thailand has been recently oriented toward well-balanced development. The "Pandintham-Pandintong" or the "Land of Virtue and Wealth" Ideology for Development of Thailand, proclaimed by the Thai Government to be the National Ideology on the May 5, 1985 on the occasion of the King's sixtieth birthday anniversary, stresses the important dimensions of spiritual and moral development in leading the social-cultural and material-economic developments toward achieving the goal of well-balanced development. This National Ideology for Development has been promoted by the Ministry of Interior in the nation-wide program for rural development about seven years ago.

The Club of Rome published a new book entitled "The First Global Revolution" (New York: Simon and Schuster, 1991) which describes the contemporary tumultuous world situation as a prelude to a new global revolution to be realized in the twenty-first century. This new revolution is shaping a *New World Society*, as different from today's society as the post-industrial revolution was from life in the preceding millennia. This global revolution is not shaped by a single ideology, but by social, economic, technological, cultural, and *ethical* factors, a sort of full-scale wholesale social change for the entire world for the first time.

The Club of Rome further foresees the need for learning a new way of life in this new era, and stresses the importance of the spiritual and ethical dimension which religions have nurtured throughout history. The fuller and more integrated development of this spiritual and ethical dimension is deemed as a necessity for a new humanism.

THAILAND'S DEVELOPMENT SCENARIO FOR 2044

It is apparent that the future of Thailand will be shaped by multiforces arising from social change which has been taking place intensely within a relatively short period of time. However, these vibrant forces can be basically classified into three categories, namely, Spiritual S&T (Science & Technology) Forces, Social S&T Forces, and Natural S&T Forces.

The Spiritual S&T Forces will primarily nurture and cultivate the Spiritual-Psychological System to a higher level of moral consciousness, and at the same time contribute to the Social-Cultural and Material-Economic System.

By the same token, the Social S&T Forces will be primarily transforming the Social-Cultural System into a more democratic stage, while radiating the ripple effects to the Spiritual-Psychological and Material-Economic System.

The Natural S&T Forces will primarily contribute to the development of the Material-Economic System toward a wealthier level, while emitting positive and negative ripple effects to the Social-Cultural and the Spiritual-Psychological System at the same time. On the other hand, the confluence of the three dynamic forces on the three systems within the next five decades will definitely bring about colossal social changes which would be far beyond our imagination.

Nevertheless, we can try our best to extrapolate the national development trend based on the presumption that future Thai Governments are seriously committed to develop the country in accordance with the well-balanced Ideology of Pandintham-Pandintong, through dynamic interplay of the three domains of science and technology.

Now, the readers will be chronologically taken on tour of Thailand which is 50 years away from the present.

THE INFLUENCE OF *SPIRITUAL SCIENCES AND TECHNOLOGY* ON THE THAI SPIRITUAL-PSYCHOLOGICAL SYSTEM UNTIL 2044

The Spiritual Sciences and Technology have been promoted in Thailand in the International University for

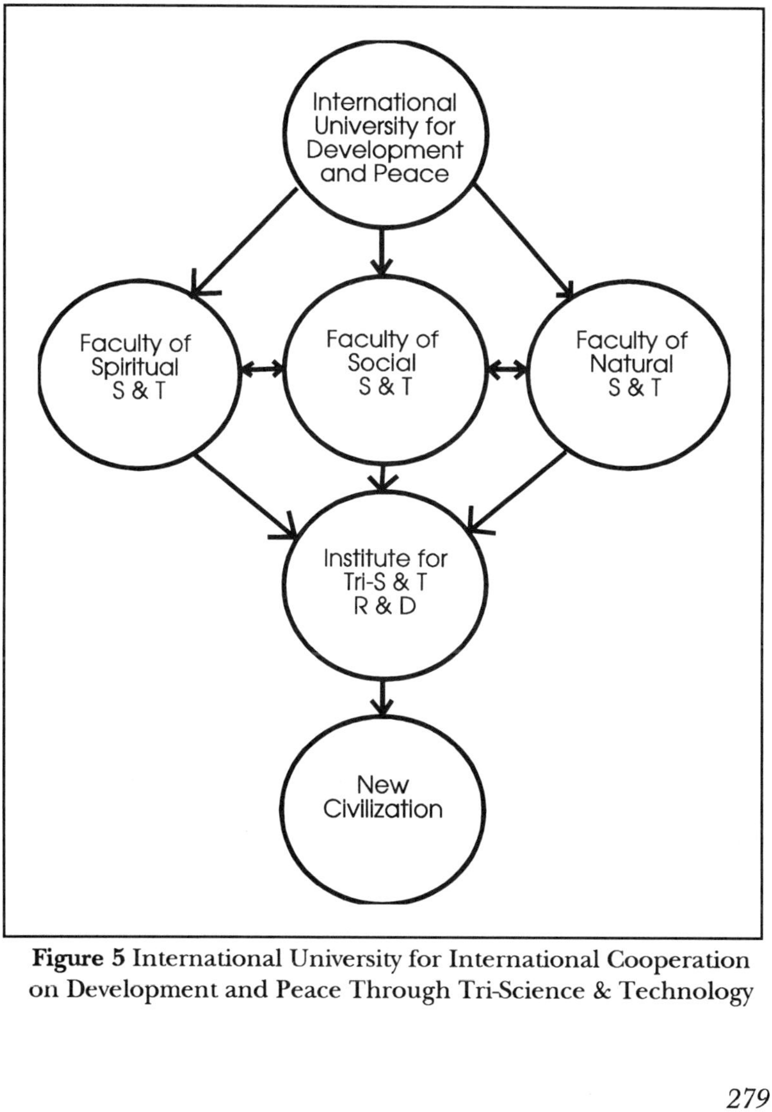

Figure 5 International University for International Cooperation on Development and Peace Through Tri-Science & Technology

Development and Peace, and in Lifelong Education System about 40 years ago. At this University, there is an International-Interdisciplinary Institute for Tri-Science and Technology Research and Development (ITRD) which has conducted hundreds of R&D Projects relating to the promotion of integrated well-balanced development and peace in the Pacific Rim Region as well as in the Global Community. Many famous international scholars have been invited to do advanced studies on several global issues and problems on development and peace.

However, the practical aspects of this Spiritual S&T had been embodied in the Pandintham-Pandintong National Ideology for Development proclaimed by the Government in 1985. The essence of this National Ideology is to promote Spiritual-Psychological Development through Religious Principles which serve as a firm foundation for Social-Cultural Development through Democratic Principles, and for Material-Economic Development through Cooperative Principles.

Within the period of five decades, Thailand has been successful in promoting Spiritual S&T to be the foundation of individual, social, political and economic developments, which render the outcome of well-balanced development and goals.

THE SHAPING INFLUENCE OF *SOCIAL SCIENCES AND TECHNOLOGY* ON THAI SOCIAL-CULTURAL SYSTEM UNTIL 2044

After four decades of the application of social scientific principles in the development of a social-cultural system in Thailand in the form of "Araya Democracy" which means "Tridimensionally Civilized Democracy," has given rise to a stable form of Thai democratic political system with well-balanced, tridimensional democratic culture.

Araya Democracy, a well-balanced and tridimensional political system has the following three dimensions:

Dimension 1: A democratic moral system serving as the foundation for every aspect of a democratic way of life. This moral system, consisting of seven natural principles for good citizens and moral principles for the social-cultural system, has been successfully inculcated in the populace through all kinds of socialization, training and education.

The seven "Natural Moral Principles" are as follows:

1. Knowing the cause: we should always be conscious and realize that any type of one's own behavior (physical, verbal and mental) may be the cause of good or bad consequences.
2. Knowing the effect or consequence—we should always be aware and realize that happiness or sadness, satisfaction or dissatisfaction of our own or of others, are consequences of our own behavior.
3. Knowing one's role and duties—we should always be aware of our status, role, duties and responsibilities and behave appropriately in that role and status.
4. Knowing the right time and occasion—we should always know when we should behave in one way that will be right and proper, and when we should not behave in another way that will not be right and proper.
5. Knowing the community—we should learn to live according to the traditions, values and needs of the community.
6. Knowing one's capacity and limitation—we should be aware of our limitations in many situations, and also should be aware of our capacity so that we would be humble, instead of showing off.
7. Knowing the individual people—we should be aware of individual differences, as well as strengths and weaknesses of other individual people, so that we know how to behave appropriately toward them."

Dimension 2: A democratic society embracing a democratic family system, a democratic school system, and a democratic political system, all of which are founded upon the democratic moral system. Through comprehensive social development in both rural and urban areas in accordance with the Pandintham-Pandintong Ideology initiated fifty years ago, Thai society now has a very strong democratic social infrastructure and institutions based on a democratic moral system which makes it possible for Thailand to achieve political stability, social peace and happiness.

Dimension 3: A democratic cooperative system representing an economic system "of the people, by the people and for the people" which is founded upon the democratic moral system and moral principles for the economic system. Thailand has been successful in developing the democratic cooperative system in terms of the National Cooperative Network and the Network for Cooperative Councils.

THE SHAPING INFLUENCE OF *NATURAL SCIENCES AND TECHNOLOGY* ON THAI SOCIETY UNTIL 2044

Thailand in the year 2044 is a full-fledged industrial country where modern high technology has made impact on all sectors of the economy.

Agriculture: Farmers who are now well-educated no longer use gasoline motor tractors in plowing and tilling the soil. Instead, they utilize the advanced solar combined tractors in doing all basic farm work from plowing to threading, milling and packaging in a single system operation.

Because of the application of advanced genetic engineering by which transgenic strains of economic plants are developed with combined high yields and pest resistance, and the advancement of agro-technology and solar-electronic irrigation system which are the backbone of modern agriculture, farmers can grow rice and other plants effectively with the average yield of five times more than what their ancestors could produce in the preceding five decades. In addition, farmers have organized a national network of comprehensive cooperative systems whereby agricultural produces are processed in the Agro-industrial Complex. Agro-industrial products are then marketed by the national network of cooperative stores who sell them to domestic consumers and export surpluses to other countries and regional economic blocs.

Through this effective hi-tech, agro-industrial and full-cycle economy, Thai farmers in the Year 2044 are quite wealthy and collectively influential in terms of political activities. They have already formed the Progressive Farmers Party (PFP) which now has the majority of seats in the Parliament and is about to form a new government for the next four years. At the local level, the PFP is very active in organizing the local Cooperative Councils which have been coordinating their efforts in the national network by promoting agricultural and agro-industrial development and business. It is also a normal practice for the Council Members to meditate for five minutes before they begin their meeting.

In summary, the modern Thai farmers are *technically effective, well-organized, and morally integrative,* thanks to the help of Tri-Technology initiated 50 years ago in the introduction of Pandintham-Pandintong National Ideology.

Industry & Commerce: Thailand is one of the advanced industrial countries with full technological applications. The major export-oriented industrial zones are the Eastern and Southern Seaboards, the Northeastern Green Belt and the Central Chao Phraya Basin. Major industries are solar cells, electric and electronic goods, computers and telecommunication equipment, plastic and agro-industrial products, gem cuttings and jewelry. Many factories have been using robots as their major work force.

The Pacific Rim Era has been well established and Thailand has been active in promoting the "Pacific Rim Economic Cooperation Council" in which Pacific Rim Industrial Development is well coordinated among the member countries and integrated with the global economy.

Thailand has effectively integrated the cooperative economic networking system with the existing free-market economy, quite similar to the Scandinavian Economic System.

"The Industrial and Commercial Management Association" which was founded 40 years ago, has been used as the main agent in diffusing *Social Technology* to the industrial and commercial sectors. The majority of companies in Thailand are now well organized and effectively managed in accordance with the Science of Management.

Spiritual Technology has been applied in many factories and companies to developing human resources in such a way that their psychological potentials are actualized for effective job performance and organizational behavioral harmony.

In addition, the International University for Development and Peace was established about 40 years ago for the study, research and application of Spiritual S&T, Social S&T, and Natural S&T with noble aims to promote well-balanced development and peace for the Pacific Rim and the Global Community.

The Service Industry: Tourism, which is part of the service industry, has grown tremendously large. More and more tourists come to visit Thailand each year to see not only the natural beauty and culture, but also the success of well-balanced development that Thailand has achieved for the last two decades.

In summary, Thai economy has been developed to reach the well-balanced goals of national development in accordance with the Pandintham-Pandintong Ideology launched about 58 years ago.

THE QUEST FOR WELL-BALANCED DEVELOPMENT ENDS WITH THE TRI-DIMENSIONAL GOALS OF LIFE AND SOCIETY.

Even though Thailand used to have a long history of well-balanced development with the Spiritual Dimension in the leading role, dating back to the Sukhothai Dynasty about 700 years ago, her modern development orientation has been derived mainly from the West and has sidetracked her toward an imbalanced development, bringing along with it undesirable social-psychological and environmental problems, ills and pollutions such as tensions, conflicts, disharmony, cultural alienation, rising crime rates, decadent human nature and poisoned physical and biological nature.

However, with the discretion of traditional wisdom, since 1985, Thailand has been trying to revert to the well-balanced model of development by proclaiming "Pandintham-Pandintong (Virtue and Wealth)" to be the National Ideology for development.

The most crucial crossroads of Thailand depends on how her succeeding governments would be seriously committed to this National Ideology by applying Tri-science and Technology to the development of the Spiritual-Psychological, Social-Cultural, and Material-Economic Systems.

It is presumed that during the course of the next 50 years, Thailand will have been able to actualize the Pandintham-Pandintong Ideology of National Development.

Spiritual S&T have mainly contributed to Pandintham-Pandintong National Ideology which employs religious and moral principles as a firm value foundation for the three systems of development. Basically, religious and moral principles have been extensively used in the development of human resources and social systems which subsequently serve as two main "Development Agents" in *appropriately and intelligently utilizing Social and Natural Sciences and Technologies for the development of the natural resources.* This approach made it possible for Thailand to consummate the tridimensional goals of life and society in the Year 2044.

Social S&T have basically contributed to the success of the implementation of tridimensional Araya Democracy which promotes a democratic social-political-economic system with moral integrity. If so accomplished, Thailand can say that within 50 years she has perfected a democratic system well-suited for the twenty-first century.

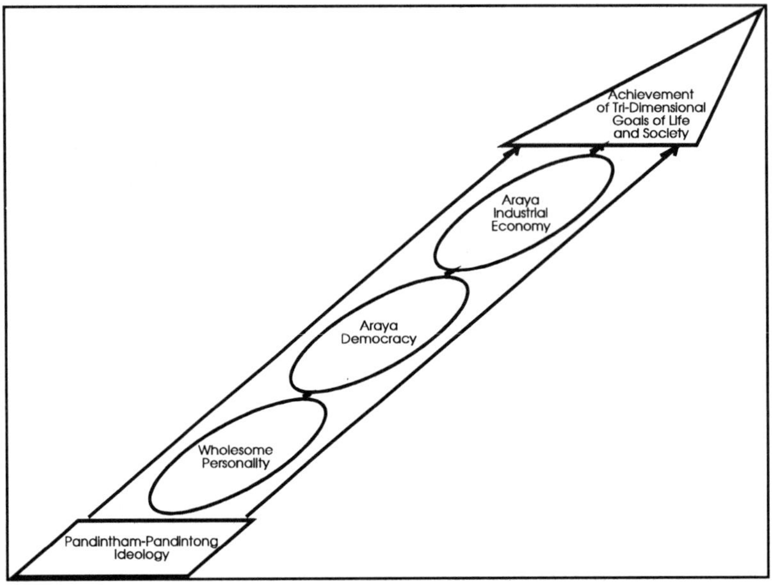

Figure 6 The Impacts of Tri-Science & Technology on Thai Life and Society

Natural S&T have fundamentally contributed to the material-economic development which renders wealth and conveniences to ultra-modern Thailand. The significant factor in this success also lies in the fact that the International University for Development and Peace has been set up for the promotion of Tri-Science & Technology for Well-balanced Development. Another important fact to be noted for the desirable fruition of Natural S&T in National Development is the interdisciplinary and integrated approach used in the planning and implementation of development projects for the duration of five decades.

REFERENCES

Club of Rome, The, *First Global Revolution.* New York: Simon and Schuster, 1991.

Ministry of Commerce, *Thailand's Exporters Selected Lists 1992-1993.* Fourth edition, 1992.

Naisbit, J., and P. Aburdene, *Mega-trends 2000.* London: Pan Books, 1990.

Wichiarajote, W., "Alternative Approach to World Development" *International Journal on World Peace,* vol. VIII, no. 3 (Sept. 1991).

TWENTY-TWO

OPPORTUNITIES FOR AFRICA

Ernest N. Emenyonu
Professor V. C. Uchendu

What distinguishes man from pigs is that men have greater control over their environment; not that they are more happy.

—W. Arthur Lewis

THE FUTURE OF FUTUROLOGY

There are two traditions of thinking about the future. The first, and the more traditional, is to evolve whatever tools of culture are needed to "discover" the future. The other, and the more modern approach, is to "create" futures with the precise tools of science and technology and educate people to accept or adapt to them.

Futurology is traditional African art. The traditional African futurologist did not create futures; he "discovered" or predicted them through inspired common sense observation. Our present handicap is that the contemporary African scholar has not seriously faced the challenge of transforming the art of his ancestors into a science-based knowledge that can truly inspire evolving technology to create desire futures.

This Congress suggests that speculations about Africa's future would yield valid development options if only we collectively try hard enough or if we could be bold and imaginative enough to turn the traditional arts of our African heritage into science-based knowledge.

There is no doubt in our minds that the future of Africa is buried in its past. Correspondingly, we cannot ignore the equally valid proposition that Africa's future is locked up with the present. Our difficulty in predicting the future, particularly in creating realistic futures for the twenty-first century, is that while we can easily build various plausible development scenarios for the twenty-first century, the facts which govern the exercise are all products of the past, and they include our scientific and technological achievements and capabilities, the men and women who created them, and the values they consider dear to them. To build realistic future scenarios demands a new concept of power and a new concept of organizational values. To create futures means much more than new technologies; it demands creating cultures and peopling society with new culture-bearers who will be able to adapt to our imposed technologies and their cultural impacts.

COMMON PROBLEMS OF FORCASTING EFFORTS

One of the major problems of all forecasting efforts is the issue of an appropriate base. A forecast, it must be recalled, is more than a prediction. A forecast falls somewhere between a prediction and a prophecy. But a forecast lacks the divine guidance of a prophecy and the empirical foundation of prediction (Bennis 1966). Given the difficulty in precisely locating its boundary, the first problem with most forecasting studies, including the American position paper for this Congress, may be illustrated by a Persian folk tale.

> There was Nasrudin, a Mulla or Islamic teacher, who lost his key. A villager saw Nasrudin searching for something on the ground.
> "What are you looking for, Mulla," the villager asked.
> "My key," said the Mulla.

> The villager went down on his knees, too, and both of them looked for the key. After a while, the villager became curious and asked the Mulla:
> "Where exactly did you drop the key?"
> "In my house," the Mulla replied.
> "Then why are you looking for it here?" asked the villager.
> "There is more light here than in my own house," answered the Mulla.

The story provides us a useful analogy in the context of our present challenge. Past and current futures and forecasting studies have tended to focus efforts where there is "light," that is, where scientific and technological breakthroughs have been achieved, not where the "keys to important problems" of the future have been misplaced or lost. Africa's monumental development challenges in the twentieth century, which are the product of past neglect, exploitation and misplaced priorities by African leaders, might benefit from the "light" of American and European efforts in the futures research and projections. It is in this context that the American position paper becomes relevant to our present discourse.

A second common problem with all forecasting efforts is the longevity of issues or aspects of issues selected for study. The temptation is to study issues that are quantifiable, and they are also issues for which data are usually easily available: Those things that are most important to man and society. Our data bank on some issues that are compellingly important today may be quite irrelevant by 2025. There is a probability that some of the issues may still be important, but we cannot predict solely from their characteristics how important they would be to society in the twenty-first century. For instance, a decade ago, nobody could have predicted the collapse of the Soviet empire or the major defaults on international debt by Latin American and African countries.

These unmeasurable events have caused monumental directional changes in the last decade of the twentieth century.

A third common problem with forecasting studies is the lack of attention to administrative structures or mechanisms needed to implement structural change in the future. The assumption that the present administrative capacity would be available or could cope if it were in place is naive. New kinds of problems projected to emerge over the next several decades demand not only new administrative systems with enhanced capacity and capability but an appropriate value-mix to cope with the new system. While we can predict the new systems, we cannot guarantee that the value-system would be appropriate or consistent with the new development.

SOCIETY IN THE TWENTY-FIRST CENTURY: OPPORTUNITIES AND DANGERS

What are the main propositions of the document which we labelled the American position paper and how do the twelve contributors to the document respond to the three common problems which seem to characterize futures research and forecasting?

The document consists of two parts: the recent technological advances and their possible utility. The technological projections are derived from major technological breakthroughs in transportation and communication and also adventures into space as a possible human habitat. The second portion of the essays consist of a number of imaginative scenarios on possible choices of living conditions on land, sea and space that might be made possible by the technological advances projected for the twenty-first century. While some of the contributors paid attention to the dynamics of human needs in predictable, utopian and dystopian scenarios, there was little or no attention to methodological issues in futures forecasting; and the growing body of research into future human needs was ignored. However, there is a strong emphasis in the document on some key processes, particularly on robotics, space and ocean habitats, and robotization with respect to their implications for the individual, society and the economy.

Hans Moravec's essay on "The Universal Robot," may be illustrative. The author makes a meta-review of robotics, describing its early uses and possible applications up to 2050. Robots are machine tools. At present, they have not achieved the status of human-like androids capable of holding a conversation. As programmable manipulators, they serve useful purposes in industry and can move tools through a pre-specified sequence of motions in hostile environments. Despite their untiring availability, predictability, reliability and precision, today's robots are usefully confined to highly structured industrial environments, and practically all their decision-making capability is pre-engineered. In labor surplus economies of Africa, the need for robots lies far in the twenty-first century, and that under specified conditions.

Hans Moravec projects an evolutionary development of robots, from Volks-robots in about 2000-2010, to reasoning robots in about 2030-2040 and hopefully by 2050, we could have the fourth generation of robots called "Mind Children."

At another spectrum of the scientific effort, and one that has the capacity to benefit humanity, irrespective of the relative strengths of national economies, are the advances in genetic engineering. Biotechnology, a discipline that investigates, develops, and applies and commercializes the variety of techniques used in manipulating the genetic material, in a wide variety of fields, is at the heart of this important enterprise. For long, it has been established that many human diseases are caused by defective genes. But little could be done for the patients until the technique for introducing replacement genes into the DNA of the proper cells of an individual that lacks the gene was perfected. The technique of genetic engineering has found increasingly wide applications in agriculture, medicine and industry. If ethical and moral considerations permit, the medical field offers great possibilities to humankind in the twenty-firt century.

For Africa, the priorities would still lie in the engineering results which biotechnology will bring to our agriculture, especially in crop protection and food preservation.

THE CHOICE OF ALTERNATIVE FUTURES

With particular reference to the architecture of the "Position Papers," one could be pardoned if the observation was made that the document achieved some structural balance. Of the twelve papers (Morton A. Kaplan contributed two papers) six were on the "softer side" of technology. The authors assumed that people in the twenty-first century will continue to be consumers of culture as they have been since they created culture. In other words, they would have a choice. If it is the role of science and technology to increase the stock of choices available in society at any given time, the ultimate responsibility to choose among available alternative futures would squarely rest on men and women. This is because we are value-laden beings.

"A Visit to Belindia" by Fred Pohl, a science fiction writer paints a scenario which may not be fiction at all for a large portion of the world in the twenty-first century. Pohl makes the point, and we agree with him, that the trends projected for robotics, biotechnology, space and underwater habitats, and in other fields, could be technologically achieved but their impacts would be far from even. He projects, without giving us reasons or justification, that the present pattern of uneven development which contours the world of the twentieth century might be carried over to the twenty-first century. Our only mild surprise is why he labels this apparent reality dystopian. Dystopia describes a hypothetical place, state or situation in which the conditions and quality of life are dreadful. We have many Belindias in the twentieth century as we had a few during the paleo-industrial period of nineteenth century Europe. Technological progress does not, everywhere, spell social progress unless there is a political will to redistribute political power and national wealth. If that challenge is not faced in the last decade of the twentieth century, it is doubtful whether even development would become a priority agenda of the leaders of the twenty-first century.

While Pohl worries about the possible intrusion into the twenty-first century of our present pattern of global inequality and individual poverty in the midst of plenty, Kaplan discusses the probability of increasing dysfunctions of our societal system and the disorganization of our personality system arising from the inability of modern society to adapt to its technological achievements. As Amitai Etzioni postulated before Kaplan, the "active society" which is a "self-transforming society," has added to the stock of dysfunctional elements it inherited from modern society: a frightening "increased capacity for macroscopic manipulation, for the generation of a sense of responsiveness where there is actually none" (Etzioni 1968: 617). While knowledge is transforming practically everywhere, its limit in social engineering must be fully understood if increased knowledge is to result in productive and wise policies. Kaplan calls our attention to some of the dysfunctions of our "scientific marvels": the political demand for increased societal control; over-population; drug addiction and nuclear blackmail by alienated groups. These problems would not only limit individual freedom but could justify the need for a "controlled society." Kaplan calls these possible developments "A Nightmare." We regard them as possibilities.

A PARADIGM FOR ALTERNATIVE FUTURES

When we think about futures and inter-futures scenarios, particularly in long term perspective, we must have human beings, thier culture and society in the forefront. Science and Technology, important as they are in building the future, do not consume the products they create. Man does, and he does so through the filters provided by his value system which is shaped, in the long run, by the available technology.

Change is a process and is endemic in the social system. Processual analysis helps us to analyze change and map its directions. Distinctions are often made between "recurrent" processes and "directional" processes of change (Uchendu 1991, 38-39). Recurrent processes of change are short-run in duration, occurring daily or seasonally, They are repetitive actions that occur in small patterns and in micro-time scales.

Directional processes, on the other hand, occur in macro-time scales. The type of changes forecast for the twenty-first century come under directional change. The sequences of directional processes are cumulative and their impact on sociocultural systems is pervasive. Directional processes sometimes produce irreversible change in the social system.

STRATEGIC MANAGEMENT MODEL

The literature on futures management suggests a number of competing, sometimes complementary methodological approaches. The method of "prospective analysis" is employed to evolve short-term strategies that are put in place to influence the behavior of projected future developments. The forecasting method uses scenarios to map the futures but does not apply intervention strategies in the present or in the future to change the possible future scenarios. It is rather a game of "blue sky" gazing, an intellectual exercise without a program of action (Lesourne & Malin 1904).

We propose a different model, drawn from managerial science. We call it the Strategic Management Model (SMM).

The challenge of today is how to manage the future. The future cannot be managed unless the present environment is understood. All top management requires some knowledge, and the more accurate the knowledge, the more effective the manager or policymaker. The top executive need not know all or possess all the skills. He or she must know where to obtain the relevant information from their organization.

Strategic management is a set of managerial decisions and actions that determine the long-run performance of an organization. It includes strategy formulation, strategy implementation and evaluation, and control. Strategic management emphasizes the monitoring and evaluation of environmental *opportunities* and *constraints* in the light of the organization's *strengths* and *weaknesses.* Following Wheel and Hunger (1972), it is important to distinguish between the *external* and the *internal* environments of an organization and their typical characteristics to assess the opportunities and challenges they pose for the future goals of the organization.

Strategic management involves three basic processes: strategy formulation; strategy implementation; and evaluation and control. For success, top management must assess the internal environment of its organization for the *strengths* and *weaknesses* it exhibits; and scan the external environment for the *opportunities* and *threats* it presents. These four strategic factors in the total environment, which are represented as SWOT (Strength, Weakness, Opportunities and Threats) must be matched, through careful analysis and assessment, if strategic management is not going to be a blind gamble. Each item of SWOT must be subjected to analysis and evaluation. Strategy formulation is the process of developing long-range plans to contain environmental threats and exploit environmental opportunities. It involves:

- The definition of the mission of the organization
- The statement of its objectives
- The outline of the required strategies
- The enunciation of the guiding policies

On the other hand, strategy implementation is the process of putting strategies and policies into action through the process of:

- Programs and programming
- Budgeting the available resources
- Procedures

Evaluation and control are the processes of monitoring an organization's activities and performances results so that *actual* performance can be compared to desired performance.

Evaluation and control must relate back to Strategy Formulation. Strategy Formulation is a function of the Internal and External Environments which produce the four major strategic factors we call SWOT. It is the relative strength of SWOT that Strategy Formulation is made, the critical factors examined being the organization's mission, objective, strategies and policies. Strategy Formulation and Strategy Implementation are interrelated, the latter is considered for its relevant programs, budgetary resources and procedures. Evaluation and control generate the standard operational procedures of the "decision rules" and provide the feedback for future Strategy Formulation or Reformulation.

AFRICAN DEVELOPMENT ENVIRONMENT

The SWOT factors identified in our strategic management model suggest that any scenarios of Africa's opportunities for development into the twenty-first century should scan Africa's external environment for the possible opportunities it has and threats it faces and understand its internal environment for the strengths and weaknesses it might reveal.

THE NATURE OF THE AFRICAN DEVELOPMENT ENVIRONMENT

Why should we strive for an understanding of African development environment? We suggest three reasons for this effort. First, it will enable us to speculate on the unthinkable: What if African economies fail to develop in the twenty-first century notwithstanding international goodwill, and imported technology? Second, it will compel us to answer the thinkable: Why is it that some African economies would realize their development potentials while others would fail to do so? Third, it will raise policy issues which must be faced in the last decade of this century if our forecasts are to be useful.

In examining the development environment in contemporary Africa, four propositions should be made at this stage.

First, development, however defined, is impossible without the "will to develop."

Second, Africa must regard the twentieth century as part of its development environment.

Third, since development has become a highly politicized phenomenon in the twentieth century, we cannot ignore the implications which African political structures and value orientations would have on future development efforts.

Fourth, continental and national realities in Africa have important consequences on the local environment.

THE EXTERNAL ENVIRONMENT

Two major influences constitute Africa's external environment in development terms: First is a century of imperial and colonial rule which created institutions and networks. Second, is the post-colonial environment with U.S.-Soviet competition and multi-national business and lending institutions.

THE INTERNAL ENVIRONMENT

Development in the African context cannot be narrowly equated with simple economic achievements. Unlike in the older states, African development involves two major political activities: the integration of a plural society and culture and the building up of a viable economy. These two realities invite attention to an important ingredient in national and continental development: the *Will to Develop*.

THE WILL TO DEVELOP

The concept of "The Will to Develop" should not be confused with "The Will to Economize."

Basically, "The Will to Develop" is rooted in the polity and is communicated through the political process to the citizenry. *It is one ingredient of development that cannot be acquired through technical assistance.* It must be homemade.

THE POLICY AS DEVELOPMENT ENVIRONMENT

"The Will to Develop" is nourished and articulated by a political system that places emphasis on development as a necessary political investment.

Four major features of the African polity which will continue to affect politics and resource development for the foreseeable future deserve attention:

(1) All but ten of the fifty independent African states have gained their independence in the last decade and a half. Given the newness of independence and the legitimate interests of the leaders who achieved independence to maintain their power, economic cooperation with near equal states is cultivated even when such cooperation is known to threaten sovereignty.

(2) "The Will to Develop," where it has been articulated is more a national rather than a regional or a continental "will."

(3) The *unevenness in development* and the varying development potential among states are probably two of the most significant facts of African development experience. This assumes geographical, territorial and national dimensions.

(4) Probably the most dramatic characteristics of the African development environment has to do with the spatial aspect of its sovereignty. *Africa has the largest number of land-locked countries of any continent.* Of the 24 land-locked countries in the world, 14 are in Africa; all of them in Black Africa.

SUMMARY

The contents of technological achievement in the first quarter of the twenty-first century are clearly discernible. For the post-industrial societies, the leaps in communication, which began in the last quarter of the twentieth century, would likely continue and the tertiary and service industries would likely dominate the labor force. For Africa, the first quarter of the twenty-first century would likely present a Belindia phenomenon but move cautiously toward optimistic scenarios as the century progresses. The key to progress would not be new technological breakthroughs as such, but would lie in organizational and political revitalization.

Unfortunately, Africa's external environment would likely provide more threats than opportunities; however, the valance could be reversed. We predict that the strengths of Africa's internal environment would be strengthened as the weakness of that environment is progressively reduced. In a nutshell, Africa would enter the twenty-first century in distress but certainly would bounce back in the century.

REFERENCES

Ayres, R.U., "The Future of Robotics" in Forti A., ed., *Scientific Forecasting*, UNESCO, 1984, 119-37.

Bennis, W.G., "Organizational Revitalization" *California Management Review*, 1966, 51-60.

Etzioni, A., "The Active Society," New York: The Free Press, 1968.

Forti, A., ed., *Scientific Forecasting*, UNESCO, 1984, 75-101.

Popper, K.A., "The Poverty of Historicism," New York: Harper Torchbooks, 1961.

Uchendu, V.C., "Towards a Strategic Concept of Culture: Implications for Continental Unity," Z.S. Ali, ed., *African Unity: The Cultural Foundations*, Lagos: Centre for Black and African Arts and Civilization, 1988, 12-25.

______., "Structure, Content and Process: A Framework for Analysis of Cultural Transition" *African Journal of Behavioral Sciences*, 1991, vol. 2, no.1, 27-42.

Wheelan, T.L. and J.D., Hunger "Introduction to Strategic Management and Business Policy," New York, 1972.

TWENTY-THREE

AUSTRALIA IN 2044

Ivor F. Vivian, Alan Barcan, and Patrick O'Flaherty

INTRODUCTION

Australia is an English-speaking democracy of seventeen million people—a population that will probably double within the next fifty years—on a vast island continent dividing the Indian Ocean from the Pacific Ocean. Its federation of six states and two mainland territories is known collectively as The Commonwealth of Australia. Marcelo Alonso (Introduction), reminds us that predicting the future of a social group, such as a nation like Australia, is "difficult, almost impossible, because of the need to include in the forecasting the unpredictable human factor, and at most we can guess... Social evolution is critically affected by ideological, political, religious and military leaders, whose appearance and subsequent actions are impossible to predict...." Even so, some previous attempts to anticipate the future of Australia have been made in recent years and we shall review these briefly.

MAKING THE FUTURE HAPPEN

The Commonwealth government established The Commission for the Future in 1986. Its Director, P. Ellyard[1] remarked in 1989, that one of the Commission's basic propositions was that "The future is something to be made, not presumed or predicted... We need to describe the kind of future we would want."

A number of recent books have considered the future of Australia. One is *Eye on Australia* by Geoffrey Blainey.[2] In another book *Furious Agreement—Forty Prominent Australians Focus on the Next Fifty Years*[3] four topics were discussed: "What do we Value?;" "Capital versus Labour;" "Learning & Expressing; and Our Place and How to Keep It." In *Our Heritage and Australia's Future*[4] fourteen eminent Australians were represented, including: Sir Charles Court; David Kemp; Dame Leonie Kramer; Geoffrey Blainey; B. A. Santamaria; Hugh Morgan; Michael James, and others. The main issues considered in this volume were: "The British heritage"; "political and civic liberties"; "the schools"; "Australia—one nation or a cluster of states?"; "Australia and the Western alliance"; "Christianity; the family; the rule of law; conservative constitutional reform"; "private enterprise"; and "economic reform." Similarly, *Crosstalk*,[5] published by the Church and Society Commission of the Australian Council of Churches, identifies ten topics relevant to the future of Australia, namely: "Peace"; "Justice"; "Environment"; "Science and Technology"; "Family Life"; "Economics"; "Community Living"; "Church and Society"; "Aborigines"; and "Politics."

Jerry Pournelle (Chapter 14) concludes with a statement that seems relevant to Australia: "any optimistic projection of the future must assume that the nation—all Western civilization—undergoes a revival of morality and finds new wellsprings of moral behavior. Indeed, one might even say that this is the very definition of an optimistic future."

To some extent the topics listed above are addressed in this paper, but an independent gaze into the crystal ball is also presented, to suggest desirable futures for Australia in 2044. The scenario tends to be optimistic and hopeful—aspiring toward Morton Kaplan's vision of "A Utopian World"—while at the same time cautioning against error, just as Kaplan has also done in "A Nightmare."

AUSTRALIA—HISTORICAL DEVELOPMENT

Until the fertile eastern coastline was discovered for the Europeans by the great explorer, Captain James Cook of the British Navy, in 1770, this vast continent was sparsely populated by less than one million nomadic Aboriginal peoples, consisting of a variety of independent tribes. Today their descendants represent less than a few percent (and "pure blood" Aboriginals less than 1 percent) of the nation. For historical and moral reasons, however, they exert a strong influence on domestic national policies, and control several, exclusive 'Aboriginal Reserves.'

In 1788 the 'First Fleet' of British colonists brought with them a small number of voluntary, professional leaders (including a Christian minister) supported by a detachment of armed marines, and a large number of convicts who were deported (mostly for quite trivial crimes) to provide a labor force. One quarter of these were women, and they were required, initially, for duties such as serving in the homesteads of the professional leaders. In this way a pioneering European settlement was established around the beautiful natural harbor of, what is now, the city of Sydney, in the state of New South Wales. It is an established fact that this colony quickly flourished, and significant also that, having worked off their 'sentence', even the convicts were allowed to own land and cultivate it. It was one of these men who—having succeeded in making a lot of money by raising Merino sheep on his property—became Pip's secret benefactor in Charles

Dickens's classic fictional novel *Great Expectations*. Australia, too, is a land of great expectations.

Australia's huge land area of 7.6M square kilometers is equal to that of the mid-continental USA, and more than thirty times that of the United Kingdom or Korea (North and South) yet its population is only one quarter that of Korea's. Australia also administers and protects a large part of the Antarctic continent—which is to Australia somewhat as Alaska is to the USA. At this time, however, political and environmental forces are inclined merely to explore and study the Antarctic rather than to develop it. By the year 2044 this may change, and we must hope that ecologically sustainable projects that will greatly benefit the world, can be created in Australian coastal and oceanic waters, including those of the Antarctic Territory, such as ocean fish farming as described and predicted by Athelstan Spilhaus (Chapter 6).

The six states of Australia, roughly in order of population and economic strength are: New South Wales (capital, Sydney), with 34 percent of the population; then Victoria (capital, Melbourne); Queensland (capital, Brisbane); South Australia (capital, Adelaide); Western Australia (capital, Perth); and the island state of Tasmania—itself the size of Taiwan. Hobart, the capital city of Tasmania, is often the last port-of-call for ships heading to the Antarctic—except whaling ships which are not welcome in any Australian ports. In fifty years time, therefore, much of Tasmania's economy may depend on the proper management of the valuable natural fisheries of the South Pacific Ocean, Southern Indian Ocean, and the Antarctic and perhaps the efficient use of Southern Ocean wind and wave energy for power generation.

Tasmania's natural beauty also will ensure that it remains a delightful place to visit, and will continue to be called "the Holiday Island." But, as a state it needs more than just tourism to flourish. It should grow also as a major center of naval architecture and ship design in Australia, while its vast timber resources must be carefully regenerated as others are being harvested. This latter task is one already performed quite well by the CSIRO's Division of Forestry and Forest Products, as described by Alan Brown et al.[6] The need for adequate food to feed this hungry world must, and can, be addressed by the Australian people in choosing their path for the future, and to that end the nation will continue its creative contributions in biotechnology. These will revolutionize agriculture to fulfill the vision of Claude A. Villee (Chapter 3) and his contemporaries.

Development in the Territories will continue to be very important. The federal government is located in the national capital city of Canberra, construction of which began in 1920, within the Australian Capital Territory (ACT). The thinly populated Northern Territory (NT)—which is that part of Australia most in need of development—is largely administered either from the city of Alice Springs made famous in the book *A Town Like Alice*—in the center of the continent near Uluru (Ayer's Rock)—or from the city of Darwin on the northwest coast. Darwin was the only town in Australia to have been attacked by repeated waves of Japanese bombers as fierce as those directed at Pearl Harbor during World War II. In 1975 it was also devastated by the cyclone (hurricane) Tracy, but it has now recovered to become a popular tourist destination. By the year 2044 Darwin is likely to become an important trading port because of its closeness to Singapore and the other Southeast Asian nations. This will be possible only if the NT road and rail transport system is greatly improved before the end of this decade.

IMMIGRATION

Many of the early colonies in America had begun in a similar way as did Australia's, but after winning their War of Independence from the British in 1776, the established American colonies rejected any further involuntary deportation of convicts and political prisoners from Britain and Ireland. Likewise, by the early 1800s the whole of the

Australian continent was developing under voluntary European settlement, and the deportation of convicts from Britain to "assist" the pioneers was no longer necessary or acceptable.

The fact that Australia was so much further away from Europe than the Americas meant that it was less attractive to most aspiring immigrants, and hence grew less quickly. Yet the smaller population, isolated in Australia, by necessity, has always been extremely resourceful, and has a very proud record of achievement in just over two hundred years. In the next century, therefore, we would expect this resourcefulness to be evident in numerous ways as Australia benefits from the best talents of many cultures, brought here to this "Lucky Country": this peaceful place; this sanctuary; this "Land of the Holy Spirit."[7]

After European settlement there were also some Chinese who found Australia an attractive place in which to seek their fortune, and today there is a significant, well-established, Chinese community within Australia. Successive waves of migration recently have resulted in a dramatic shift from an almost exclusively "White Australia" policy to a so-called "Multicultural" policy. The latter is not without criticism, however, since some ethnic problems are beginning to show themselves in community resentment toward new "migrants" during times of high unemployment, and the economic, cultural, and linguistic problems that each new wave of immigrants bring into Australian domestic affairs.

Currently, Australia has accepted settlers from 120 countries. The largest overseas born group, comprising 1,208,300 or 7.2 percent of the population by 1989, came from the United Kingdom and Ireland. The second largest group are those born in Italy comprising 267,600, or 1.6 percent of the population. The proportion of Australian citizens born in Asia has increased from 3 percent in 1947 to 4.3 percent in 1989. The largest Asian-born groups established in Australia are: Vietnamese (110,000); Malaysian

(70,000); Filipino (60,000); Indian (57,000); Mainland Chinese (50,000); Hong Kong Chinese (47,000) and about 40,000 Koreans. Clearly, as a result of Australia's active involvement in the major wars of this century, in the last fifty years particularly, many of the world's oppressed peoples have come to Australia seeking peace, a safe refuge, and a better life. Following the war in Korea and Vietnam, especially, there are now significant numbers of Koreans and Vietnamese in Australia, not to mention all other Asian nationalities. Many also came simply to escape from communism and other forms of tyranny. Australia also has a growing number of migrants from the USA.

The need for Australians to accept cultural diversity in immigration is far more than mere tolerance. It evokes a need to use cultural differences in a creative, life-enhancing way in the interests of society as a whole. There is also, however, a need to define a desirable culture for Australia to aspire toward in the next fifty years, that is both virtuous and free from the injustices of old world cultures—such as the caste system of India, or the abuse of women under some religions.

Although the rate of population growth in Australia is relatively high compared with that in developing countries, it is below the growth rate for less developed countries, such as Indonesia (2.0 percent), Iran (3.2 percent), Vietnam (2.2 percent) or India. Australia's *per capita* intake for third country resettlement, on the other hand, is one of the highest in the world. The magnitude of the international flow of refugees, however, is well beyond Australia's economic and social capacity to absorb continuously at the present rate. Conversely, humanitarian considerations for the plight of genuine refugees has swayed immigration policy and is the reason why there are now many people in Australia who cannot speak English—much to the growing displeasure of the established community.

An aspect of current immigration policy that will have an impact on our future is multiculturalism, and the "jury" is still considering its verdict on the long-term effects of such a policy. Opening the most recent annual conference of the Federation of Ethnic Communities Councils of Australia, held in Sydney in August 1991, the Governor General of Australia, the Honorable Bill Hayden, who has been a strong advocate of this policy, commented that "if we stop immigration today, Australia will continue to have one of the most multicultural societies in the world" (although some people regard this as culturally undesirable for this nation). And he added "we must service this ethnic diversity by means of programs and services across the spectrum of Australian society in order to move closer toward the ideal that all Australians have equal opportunity and equal access at the economic, social and political level."

A key element in "servicing" our ethnic diversity is an "enlightened" language policy which, in Hayden's Utopia, presumes that Australians are willing to conduct their affairs in a multiplicity of languages. It is a policy that becomes increasingly unacceptable to many Australians, and is greatly in need of revision. Some say that multiculturalism is good because it recognizes that the future world will involve increased global interdependence. Already more than 15 percent of Australians use a language other than English at home.

Conversely, opponents of multiculturalism argue that it is not reasonable for a nation which once spoke a common language to be so distorted by social engineering that it is unable to function properly without enormous government expense, printing every document in a dozen languages, and trying to keep every ethnic whim satisfied. This view suggests that there needs to be more coherence in Australian culture, not less. A standard and level of culture should be aimed for that is better than that of any one cultural group that has entered Australia so far, rather than a hodgepodge of competing ethnic groups that at one time migrated there. On the other hand, there is a need for an ethnically-diverse TV channel, such as one already in existence, which has fulfilled a useful role.

CULTURAL AND ETHNIC PROJECTIONS: THE REPUBLIC OF AUSTRALASIA IN 2044

In the immediate future there is an urgent need to ensure an appropriate infrastructure to cushion the settlement and integration process, and provide training in the English language for those who cannot speak it. At the moment, for example, our most popular migration destination, Sydney, faces enormous infrastructure costs, and this is compounded by the large number of recent migrants who cannot find work, because the community is beginning to reject them from the workforce, especially when they cannot speak our language.

As with the general population, so also the ethnic communities are aging. Hence, the future society will see an increased number of retirement villages which will cater to the special needs of their members during their twilight years. On the other hand, well-educated, second-generation, Asian Australians and Eurasians are excelling at schools and universities, and their impact upon the less motivated, easy-going Europeans will be felt in the future as Asian Australians take more top jobs in the community. Some, like the late Dr. Victor Chang, Sydney heart surgeon (murdered in 1991 by Indo-Malaysian gangsters who entered Australia to extort money from him) have become famous and highly respected already.

It is likely that in the future the policy and administrative arrangements for the selection, acceptance, arrival, and settlement of migrants and refugees will be undertaken by a Department for Population and Local Government, rather than by the present Department of Immigration and Local Affairs—thus reflecting the fact that migration policy is the one controlling variable in population control. In the long-run, a population boost is expected to stimulate economic growth through the increased demand for housing, goods, and services, and to inject fresh skills, capital and enterprise. Present economic arguments about the value and cost of migration and the settlement of refugees, however, seem unsure and inconclusive. The quest for economic stability, and environmental conservation, in the face of very high unemployment levels, must impact on the overall immigration intake.

Australia has some degree of social cohesion which offers hope to the global community. Our nation is not yet marked by the systematic and organized racial violence experienced in other countries, but it could become so if some current trends continue unchecked. A vision of the future involves continued application of the principles of social justice to assist immigrants and refugees to reestablish their lives in an environment of mutual respect and tolerance. The right to express one's own culture, beliefs and values is a reciprocal one, but at the same time newcomers must be shown a vision of what our Australian community and culture is trying to achieve and be encouraged to contribute to it in a coherent way. The tendency to create isolated blocs or even ghettos must be discouraged.

A major feature of contemporary Australia, therefore, is the considerable diversity of culture and ethnic mix, particularly in the major cities. By 2044 it is inevitable that Australia will have greatly strengthened its links with all of East Asia, and the nations of the Pacific Basin in particular. It is possible, and some even say desirable, that in the next fifty years Australia's multiracial communities will intermarry and establish an embryo Eurasian society. In the future, the ethnic composition will have changed significantly. An identifiable, highly-virtuous, and coherent Australian culture must be established, and all racial prejudices need to be overcome by education rather than law.

Inevitably Australia's trade, growth, and immigration will develop from its Pan Pacific connections. Many South Pacific Island people have migrated to Australia (where there will soon be more New Zealanders than there are in New Zealand) and in most cases they are seeking greater opportunity. It is probably desirable, therefore, that Australia, New Zealand, and all the South Pacific Island nations in the Oceania Region should, before the year 2044, form an extended Commonwealth or Republic. Let us call it "The Republic of Australasia"—assuming that it will include New Zealand.

In The Republic of Australasia each nation would be a separate state and have its own local government, but under a central Federal System that was constituted to protect, develop, and benefit all of them fairly—in much the same way that the states of Australia are linked to their Federal government in Canberra. Community migration between and within these member states of Australasia would be completely free, and of course they should also adopt a common currency: the Australasian dollar; and a common language, which should be English in the first instance, but there must be a carefully-planned priority second language policy. In education the timescale of effective planning is many decades, and its implications can be felt for a century or more. The educational policies of the Republic of Australasia must be planned very carefully.

EDUCATION

For more than a hundred years, until the early 1970s, education in Australia was the responsibility of the various states. These maintained strong centralized systems of state schools. Alongside these, a separate, decentralized Catholic system exists, as well as some corporate Church or independent colleges. In recent years this tradition has been diluted in favor of much greater decentralization of state-school systems and of the educational bureaucracy. These reforms represent a massive shift in policy—the essential features of which are likely to survive the next 50 years—for better or worse.

The current reforms are a response to several major forces. The economic crisis, and high youth unemployment, has emphasized the importance of vocational studies and is producing a rise in the number of adolescents persisting in the school system beyond the minimum required age (15 in most states) partly to improve their skills and partly to avoid unemployment lines. The spiritual philosophical crisis had already undermined the strength of the former humanist studies. The long-established curriculum[8] was replaced in the early 1970s by one focused on individual personal development but also on the concerns of the special interest groups prevalent in a pluralist society—content was depreciated, while process was emphasized.

Politicians have now become dismayed at the quality of education produced over the previous two or three decades by state schools which had been under the control of the "education professionals." Some militant teachers' unions also had a heavy input into education policy and practice. Politicians have attempted to restrict control over the curriculum which the schools and teachers obtained in the later 1960s, and have sponsored attempts to develop a national curriculum and to set basic standards. They have also attempted to devolve power to individual schools (which could be more economic and reduce the influence of the teachers' unions). Another aspect of reform has been attempts to reduce the size of the educational bureaucracy, and to decentralism its structure, in parallel with decentralization of control over the schools.[9]

EDUCATION: WHAT IS WANTED

Australian education has a number of strengths. These include the generous provision by the state of material equipment (buildings, etc.) and personnel (e.g., small class size because of a generous pupil/teacher ration of 28:1) in state schools, generous funding by the state of non-state schools and widespread public awareness of the importance of education and training. Another strength is the development of the method of selling higher education to Asians, as evidenced by the large number of young Asians studying in Australia. This is possible because English is an international language, and Australia is the closest major English-speaking country to the east of Asia that also offers ready access to higher education. The state schools should teach only the most relevant Asian languages and one of these predominantly, as the priority, second language. Arguably Korean should be the high-priority second language, then Japanese, and German the next, because:

(a) the culture of the nation of Korea, acquired over 4325 years of recorded history, is highly developed and has much to offer the Christian culture of the majority of Australians, more so than most other Asian cultures. Korea was identified by the Evangelist Billy Graham as having the world's most devout Christian community. A significant number of Koreans are Christians—as are most Koreans who have settled in Australia—the rest being Buddhist or Confucian;

(b) though it is not obvious yet—as we discuss in our section on economic projections—it is predictable that, in the next fifty years, North and South Korea (combined) will become both the dominant economic nation in Asia—eventually overtaking Japan—and also an industrial giant that will dominate the world economy in much the same way that Japan does at the present time. For this reason, Korea will become Australasia's largest trading partner, overtaking Japan. It will be extremely important for the cultural and economic development of "The Republic of Australasia," therefore, that as many citizens as possible can speak, read, and write the Korean language. Such considerations have led to the development of "Korean Studies" in several of our universities, and to an interest in fostering research and study links between Korean and Australian universities.

(c) in its written form Korean (Hangul) is one of the few oriental languages that has a phonetic alphabet similar to the European languages (24 basic letters in fact) with vowels and consonants etcetera, but grammatically similar to German. For this reason, and because of the dominance of Germany in the European culture and economy, and because of its importance in science and engineering, it is also appropriate that German should be the dominant European second language taught in Australian state schools—more so than Italian or French, etcetera, which can be taught by the private schools, as is necessary or desired. Some already teach Latin.

One of the weaknesses of the present system of education in Australia is the gradual replacement of the old tradition of "fair average standards" by a new one leading to extremes of excellence and depression. The shift in the determination of educational policy from the educational professionals and teachers unions to the politicians and non-educational administration bureaucracy has, because of the educational inexperience of the politicians and lay administrators, and the resilience of the educational bureaucracy, opened the way for the recycling of exhausted theories and old educational placebos which have been tried and proved futile in years past.

A conflict exists between pressure for utilitarian skills-oriented education and general education in the schools. The same blurring has occurred at the post-secondary level, in the new universities created by the recent mergers of staid traditional universities (which once catered to the professions) and Colleges of Advanced Education, which were intended to be vocational but in many cases developed into dynamic and innovative institutions. The subsequent mergers have created some 34 universities of varying sizes (mostly large) and offering differing curricula, but under strong federal government influence. Academics have previously played a reduced role in Australian social and political life but this nation is just beginning to discover the value of higher education. In recent years, therefore, such education has become necessary in order to enter a range of "white collar," professional, occupations.

Pressure has mounted in the last two decades to turn schools into social welfare agencies and, more recently, into vocational training institutions. For many young, potentially unemployed, people they are becoming holding pens, whose function is both social (custodial) and educational (vocational training). Current proposals seek to ensure that the vast majority of Australian youth will be kept off the unemployment lines by semi-compulsory attendance at educational or training institutions. The prolongation of the period of formal education means that many young men and women will not have completed their basic education until they are in their early twenties, which is penalizing nonacademic teenagers. Attempts to introduce competency-based learning for employment preparation, involving industry, unions, and government, offers a tentative solution.

An aid to intelligent guesswork about the future is to study the lessons of history. Epochs in the development of civilization which resemble contemporary culture will provide hints and offer suggestive parallels. The most obvious historical lesson is from the Hellenistic Age, which developed after the conquests of Alexander the Great united the societies of the Eastern Mediterranean and Western Asia. This culture was taken over by the Roman Empire, another great synthesizer, in the first century AD Analogies between the Hellenistic/Byzantine society and that of contemporary Australia are apparent in the prevalence of: cosmopolitanism; multiculturalism; pluralism and a commitment to formal education; womens' liberation and feminism; falling birthrate; widespread philosophical doubt; a gap between popular culture and high culture; a sense of insecurity; family instability; and uncertainty about the future.

In our secondary schools this fascination for what the future holds has taken the form of "Futures Education." This is usually a segment in the English curriculum. It operates through the study and discussion of utopian, doomsday, and science fiction novels—for example: students in New South Wales in 1990 studied Aldous Huxley's *Brave New World,* Ray Bradbury's *Fahrenheit 451,* Rachel Carson's, *Silent Spring,* or George Orwell's *1984.*

The strong movement currently underway to decentralize the administration of education and to devolve more powers on the schools could result in community control of schools, so that the distinction between state and private schools is diminished. We could see a variety: religious-based schools; ethnic schools; parent-controlled schools; and state-subsidized schools for the handicapped. Furthermore, limiting the size of schools is vital for a religious-based or humanist curriculum and for individual attention to pupils. A possible optimum number would be 800 pupils for primary and secondary schools. In such small schools it is possible for staff to know each other and to know the children.

Confusion has grown over the role of the schools. Are they for: social adjustment? development of technical skills and vocational preparation? or broad, general academic education? The answer is that we need all of these, but in what proportions, and at what standards? One scenario suggests that education should be recognized as a function and responsibility of family, school, universities, the media —especially television—and also adult education colleges. We must also use computers more effectively in the process of educating all age groups.

The Family

There is a need for closer contact between schools and families in a variety of forms, such as school councils; home tuition; increased parental support for the moral education of children; increased parental assistance with academic knowledge which well-educated parents can provide in the home; the provision of motivation; and emotional support for the children in their studies. The choice of schools for parents' and students' sake can be enhanced through a voucher system.

The Schools

In the schools the curriculum is the main problem. Primary schools have been neglected for the last 25 years. The need for a clarified syllabus and agreed standards is now desperate. In the long-term, what is needed is the development of basic skills plus nonsecular, humanist, academic studies. In secondary schools, one great need is for a division between junior secondary (ages 7-10) and senior secondary (ages 11-12). This already exists in some Australian systems. Such a division can provide smaller schools with a more specialized curriculum, greater unity of purpose within the institutions, better discipline, and greater economic efficiency.

The Universities

Apart from an optimum size of about 4,000 students, which allows academic staff to know each other and their students better, universities also need to be specialized. Some universities should be primarily liberal arts ones, some technical, some commercial, some devoted to the tertiary or service vocations. Specialization gives a unity to the staff. It solves the problem of selection of students, by encouraging self-selection. Specialization is likely to raise standards and is not narrowing in outlook if it is based on a preceding sound broad (nonsecular) humanist primary and junior secondary schooling.

The Media

We must strive to diminish the consumerism and commercial emphasis in television by fostering competitive, autonomous, community-funded stations. The vital role which television plays in moral education and dissemination of values requires a firm censorship of films and videos, to eradicate pornography and the graphic depiction of violence. Magazines and newspapers should be less censored, because freedom and independence of the press is important in a democracy.

Adult Education

Education of mature students is provided in part by the media while organized adult education classes also exist at such institutions as evening colleges, technical colleges, the Workers' Educational Association (mainly in NSW) and, more recently, the University of the Third Age (for adults over the age of 55). Television as a form of adult education is not specific to a particular audience—its audience includes children watching "adult" shows. The participants are passive rather than active. Nor does television provide a cumulative, articulated succession of studies, although attempts to provide some "open" courses via TV (as in the UK) have been initiated.

It is important to attempt to revive formal liberal adult education, as an alternative to television, for groups rather than individuals. This suggests work-based education, which also benefits society because it helps shape attitudes. Forms of adult education should vary according to the different groups: workplace groups; rural landholding groups; mature-age urban groups. For workplace groups the emphasis should be both on their own vocation, and cooperative development.

The Curriculum

At all levels in schools, universities and adult education, the curriculum is vital. More broadly, the development of a moral-spiritual-philosophical education is vital. A useful base is to develop a sense of cultural heritage, built on the recorded human experience of the last 5,000 years, so that this nation can aspire toward a higher culture of its own. This can be promoted through the use of special videos such as

Kenneth Clarke's "Civilization" series in the late 1960s which provided an early model, and textbooks that need to be specially written and made available at low cost. In some cases, the reprinting of late nineteenth and early twentieth century textbooks might prove valuable. The development of international understanding will be an important consideration in the next 50 years, so the curriculum needs to encourage a proficiency in important languages.

ECONOMIC DEVELOPMENT

Presently in Australia the two major political parties are the Labor Party which has a socialist ethic and the more conservative Liberal-National Coalition. There is currently little difference between the major parties except in emphasis. A small Australian Democrat Party, The Call To Australia Party, the Green Movement, and a number of influential independents also hold the balance of power in a few states, and have toppled more than one state government so far.

State intervention in many aspects of daily life has been a long tradition in Australia and an edifice of Social Welfare was created early in the twentieth century. Today the welfare state is in trouble—beset by many economic difficulties. The 1980s were marked by deteriorating economic conditions, creating a recent growth in unemployment to about one million, and a prolonged, accumulating increase in the balance of payment deficit to more than A$1.3 billion.

Fifty years ago Australia realized, for the first time, that its future and prosperity might not lie so much with the British Empire and Europe, but with herself, with her great ally America, and even with the troubled region of Asia. Social and ideological values were secure, while secondary (industrial) production exceeded the value of primary production for the first time in Australia's history. Today, the crisis is economic, but social and ideological values are also in considerable confusion. The military challenge of 1942, the expansion of a dominant imperial power across East Asia, has been transformed into a peaceful challenge of the economic progress of many countries of East Asia. Alongside primary and secondary industry, a new "tertiary" or service industry has developed. Australia must improve its performance economically, socially, culturally.

Developments to our north will be highly relevant to the shaping of our own future. It is safe to assume that in fifty years' time China will have abandoned communism, and its economy will be developing rapidly with the aid of Korean technical assistance (more so than Japanese). Consequently, both Korea and China will begin to outshine even the economic miracle of Japan. China, at present, only takes about 3 percent of Australia's exports, though Korea takes much more; and this is likely to grow markedly in the next half-century. Yet with such large internal reserves of space, materials, and population, China is likely to be inward-looking until a modern industrial base is established with help from Korea.

Korea will be united and a strong influence on world economic affairs and is therefore likely to be highly influential for Australasia. Korea, more so than Japan, will be sought after to help in the development of Siberian Russia, the reindustrialization of the CIS, and of course, China. In fifty years' time Korea will become an industrial and technological giant that will require huge quantities of Australian minerals and other produce. Trade with Korea will soon exceed that of Japan and Korea will become Australia's most important trading partner

Japan will continue to have a strong economic and political influence on Australia, as well as having considerable international strength. At present, Japan takes around 27 percent of our total exports and is an important source of capital investment in Australia—as well as sending strong yearly contingents of tourists.

Malaysia was once one of the few overseas countries in which Australian capital was invested. At the moment,

Australia is an important source of higher education for Malaysian students, and such trends will continue no doubt into the next millennium. Unfortunately, however, there is a subtle hostility toward Australia which precludes, at present, a long-term relationship of trust developing.

Indonesia will probably have an expanding economy. Its population is likely to reach a peak of 230 million by 2044, but Australian links with that country will remain strained until its military leadership is reformed. Until then, despite sterling efforts by the present Prime Minister to improve relations with Indonesia, many Australians remain wary of Indonesian government motives, and trade with that country is not likely to be so vital for Australia for some time to come. The culture of Indonesia, which is Moslem, is vastly different from that evolving in Australasia. For this reason immigration from Indonesia may be naturally limited.

OPPORTUNITIES AND DANGERS

A number of already existing strengths that will help Australia in the future are:

(1) We attract an increasing number of visitors; the growth in the tourist industry has brought considerable benefits, financial and otherwise. We lack old-world historical relics, but we have interesting and distinctive flora and fauna.

(2) We are a vast land with important reserves of minerals, including uranium ore, which will continue to be significant for the national and world economy.

(3) We are geographically isolated, but this has brought protection from exotic diseases (though smuggling of people or plants into the country might endanger this). Our climate encourages good health, an appreciation of sports, and open-air activity.

(4) We are still able to offer fair average standards in education, though this has been eroded in recent decades. Nonetheless, quality education is still to be found and attempts are being made to restore these standards nationally.

The attractiveness of Australian education is evidenced in the sale of higher education abroad, particularly to East Asia.

(5) We are a socially stable society, based on a strong spirit of goodwill, tolerance, and relaxed relationships. The old Australian tradition of "mateship" survives. Class divisions have never been very rigid, and social mobility is possible.

On the other hand, Australia has many weaknesses too. One feature of the economic and social crisis is the burgeoning of the external debt. At the end of 1990 this stood at A$134.5 billion, or 36.2 percent of the Gross Domestic Product for that year. At the end of 1991 it had risen to $144.7 billion, 38.2 percent of the GDP. By contrast, our neighbors Japan, Korea, Taiwan and Singapore have accumulated massive foreign reserves. High unemployment (especially of young people) currently well above 10 percent of the workforce, is another sign of malaise.

There is over-reliance on aspects of the economy which cannot bear the burden (rural and industrial) coupled with neglect of "service" or tertiary industries and export-oriented activities. Among individual Australians there is a considerable dependence on credit cards to maintain living standards or simply to survive, because the banks currently require most loans to be in this high-interest format.

For some ten years Australia has maintained a standard of living which the economy cannot support. The cost of the welfare state has risen. Lacking the resources to sustain the current lifestyle, Australia has had to borrow from abroad, sell segments of the economy to overseas investors, attract overseas capital by offering high interest rates. Privatization of state-owned enterprises is forced on governments which can no longer sustain the level of social welfare payments without providing new sources of income other than taxation.

Financial cuts plague our educational and health systems. The economic crisis has implications for our immigration policy (which we have discussed already) and has weakened family life and stability in several ways: single parent families;

a high divorce rate; increased cohabitation without marriage; widespread alienation of children from their parents and so forth. There is confusion over the role of the sexes. Moral and economic problems have generated an increase in crime. In a contracting economy the possibility of ethnic clashes is not to be discounted.

RELIGION AND SPIRITUAL LIFE

In defining at least one aspect of existing national culture we can certainly say that Australia is predominantly a Judeo-Christian nation. About 26 percent of the population is Catholic, 25 percent Anglican, and the remaining 49 percent include the other Protestant Christians such as the Presbyterians and Baptists, the Uniting Church (which was formed in 1977 by a union of Methodists, Congregationalists and some Presbyterians), Greek and Russian Orthodox, Jewish, Buddhist, Moslem, Hindu, and all other religious minorities, and of course some who have no religion at all. In recent years the number of Buddhists, Moslems and Hindus has risen rapidly because of the change in immigration policy toward multiculturalism.

In considering the role of the spiritual in Australian life today, one can note with Archbishop Hollingsworth that, while attendance at formal services has declined, and understanding of doctrine has become somewhat confused, more than 80 percent of Australians still believe in God. The eminent historian, Dr. John Eddy, SJ, in a recent conversation suggested that we are now coming to an end of the secularist era, which is the last ember of the conflagration of ideas associated with the French Revolution. The recent events in Eastern Europe and the former Soviet Union are evidence of this, and Australia will be affected by these events for the next decade or so. Dr. Eddy believes that major historical events cause disturbances which take at least three or four decades to settle.

Dr. Eddy argues that there are two tides flowing within the religious community of Australia. The first is a growing recognition of the role of the spiritual, leading to interfaith cooperation, harmonious relationships, recognition of the indigenous spiritual values of the Aborigines, and the new insights of recent arrivals with Eastern religious traditions, and culture.

The second is a reaction to uncertainty and rapid change which evokes the certainty of fundamentalism, in each of the major world faith groups in Australia. Historically, Australians have been afraid of, or suspicious of, the "stranger": (which was exemplified in the fearful and defunct "White Australia Policy"—with its xenophobic fears of the "Yellow Peril" from the north) and this tendency, combined with fundamentalism, can lead to a disastrous fragmentation of our society, even if it is substantially "religious" and for the most part "tolerant."

Professor Max Charlesworth like the American theologian John Hick[10], considers that the tendency to pluralistic understandings bodes well for Australia. In his paper, "2000 AD: Terra Australis and the Holy Spirit," he envisages a new spirituality emerging in Australia from Aboriginal spirituality, the greater cooperation and understanding between Australia's mainstream religious groups, and, possibly one can add, the natural spirituality of the environmentalists who seek to care for and love the whole of creation. Professor Charlesworth envisages Australia as a center of pilgrimage, a spiritual crucible, as was India some 3,000 years ago, or the Middle East 2,000 years ago.

The disenchantment of many Australians with traditional religious structures is highlighted in an article on church and society by Sister Veronica Brady in *Crosstalk.* In *The Sunburnt Soul,*[11] Dr. David Millikan points to the growth of religious communities outside normal church structures, known as House Churches. This supports Dr. Eddy's view that Australians are seeking new spiritual goals but still being wary of "professional" religious leaders. The "Green" movement is certainly spiritual in many of its aims and promoted values. The late eminent historian Professor Manning Clark posed the question: "How can we feel a reverence for the whole of creation without... a religious view of the world?"

The interest in religious matters by the Australian public and even its media is evidenced by public reaction to criticism of politicians, and their promises on social reforms, by Archbishop Hollingsworth who believes that the axioms which need to be pursued to achieve a better society are those of justice, sustainability, participation, and efficiency. He has had strong community support and respect for the stand he has taken. His views support our contention that values derived from spiritual thought must help shape the society we require in the twenty-first century. The community also takes a strong interest in the internal controversy of the Anglican Church on the ordination of women to the priesthood. The issue of what ordination means and its biblical authority have become major items of discussion in the popular press and other media.

If Eddy, Charlesworth, and Hick are correct in their view, then leaders in Australasia will need to reconsider the role of religion and spiritual values in society, and to promote the tolerant and cooperative spirit that will result in a pluralistic understanding of the major faiths, so that educated Australian Christians will gain deeper insights into their own tradition by considering the special insights of other world faiths within Australia, and of our own indigenous peoples. An ecumenical sharing of ministry between all believers, and involvement in worship with each other (i.e., all people of goodwill) will bring about Charlesworth's dream of *Terra Australis—the Land of the Holy Spirit.*

THE ENVIRONMENT: A MORAL CHOICE

In considering what would be the desirable state of Australia's environment in fifty years' time, many researchers and environmental analysts throughout Australia (and, indeed, throughout the world) are conscious of the need for "Sustainable Development."[12] While many definitions of this term abound, depending on whether the view is conservative or business oriented, currently the most generally accepted definition is: "sustainability as a value—a moral choice of accepting intergenerational equity as an overriding ethic."[13] The most critical aspect of sustainability, in the long-term, "is not economic variables *per se,* but rather the ecological entities and relationships."[14]

Despite its geographical isolation and overall low population density, Australia has not escaped many of the global environmental ills, such as ozone layer depletion and the greenhouse effect. In addressing the future of the Australian environment, however, we must note a number of unique factors. The geological age of Australia is reflected in the low nutritional value of its soils, while Australia's geography renders much of the climate unsuitable (either seasonal or marginal) and much of the terrain unfavorable for sustainable human development. The most serious problems facing Australians, as they contemplate the state of their native land 50 years from now, are arguably those of land and water degradation.

Specifically, in regard to water supply, Australia only receives the same amount of water run-off as much smaller countries like New Zealand or Vietnam, and receives somewhat less than Japan. Currently only 4.5 percent of this run-off is regulated and the maximum potential regulation of this run-off is only about 25 percent. This is because a large proportion of the rainfall occurs away from population centers. The island state of Tasmania, for instance, receives 10 percent of the national rainfall but only 2.7 percent of the population lives there.

Northern Australia, which is sparsely populated, also receives a major portion of the national rainfall but, after soil and terrain considerations are taken into account, contributes only 1 percent to the available arable land for development (down from 31 percent after climate considerations only). Concerning water consumption in Australia, 75 percent is used for agriculture, the majority of which (70 percent) is dedicated to irrigation. Half of the water irrigation is for pasture, which is a low return crop. In considering how to achieve the economies that will maintain sustainable supplies of Australia's sparse water resources, careful consideration must be given to the manner in which it is used.

Water quality is an even more serious problem than water supply. Sydney's water, for example, exceeds the World Health Organization's guideline for level of suspended solids. It is likely that, in fifty years' time, there will have been a decentralization of population due to the pressure on the large urban centers like Sydney, with their associated infrastructures which are barely coping with the current nationwide population of 17 million, let alone the projected national population of 26 million in 2031.

In terms of land, Australia remains unoccupied to the level of 30 percent. Less than 10 percent of the total surface area of Australia, at the moment, can be considered suitable for growing crops or sown pasture. Two-thirds of this has already been developed. The remainder is mostly located in Eastern Australia (18 million hectares) and is more marginal, so it will require a high level of management to develop sustainably. Of paramount importance in the next 50 years, however, will be the management of coastal areas, since 98 percent of the urban population of Australia lives within 20 kilometers of the coast. This compares with 41 percent for the USA, 30 percent for Brazil, and less than 5 percent for China.

Land management practices adopted since the time of European settlement have taken their toll of both the productive capacity of the land and the native flora and fauna. A key factor has been the extensive land clearing and overgrazing (Australia has more than 150 million sheep and goats) which has led to habitat destruction, loss of native vegetation and soil erosion.

In order to achieve a desirable state of the environment in fifty years' time, the current trends of lack of sensitivity and irresponsible stewardship of the natural riches of the continent must be replaced by what Professor Henry Nix of the Center for Resource and Environmental Studies (CRES) at the Australian National University, has termed "a conservation ethic."[15] He calls for a more caring and responsible attitude to other humans, other life-forms, and the environment as the basis for all other initiatives to ameliorate the degraded state of Australia's natural resources. Efficient use must be made of resources and excessive consumption must be avoided. Professor David Yencken of the University of Melbourne adds that: "In order for the human to modify its collective behavior to ensure sustainable development, changes will be required in legislation, in government practice and public culture, in economic planning and national accounting, in business culture and practice... as well as in personal behavior."[16] He remarks that the processes and ingredients for such a change are immensely complex. It could be argued that failure to address the environmental problems facing Australia would present the greatest threat to peace on the island continent. Professor Moore believes that appropriate pricing of natural resources is one of the key issues in the resolution of conflicts between resource use and conservation.

TECHNOLOGY OPTIONS

In Australia the project of highest priority at the moment is the improvement of roads and rail transport in all states so that by the year 2044 there should be a standardized, high-speed compatible, rail network around the entire country. This should be backed up by a freeway system that goes all the way around the country and through its center. There has been a plan to create a very fast train line from Sydney to Melbourne, but it lacks adequate funding at the present.

In Australia, where distances are vast, and air travel is a major transportation method (and has an excellent safety record) there must be faster, safer, and less costly modes of road and rail transport between all major cities. Railroads are little used for intercity travel in Australia because the present services are much too slow (though they were once the major means of transport) but the quality of some interstate roads is so poor that bus accidents have become too common —which has caused great public concern.

Currently, there are very good roads in some areas, but there is still no continuous high-quality freeway up the Eastern coastline from Sydney to Brisbane, for example. High speed safe roads are woefully lacking for long distance travel and delivery of goods. The supertrain concept should be developed as soon as possible, and extended to link all of the major state capitals: especially the Melbourne-Canberra-Sydney-Brisbane-Cairns route along the Eastern coastline, which is the most populous and highly-developed part of Australia. Remnants of the failed multigauge rail system must be upgraded to a common standard that is best for the high-speed trains, even if, in the short-term this is more expensive. The high-speed train system should also link Sydney with Adelaide, then Perth or Alice Springs, Darwin, and eventually link up with Cairns and the Cape York Peninsular, in N.E. Queensland.

It is in the Northeastern tip of Australia that very high-technology development should take place. There is a plan that a new spaceport should be built there with the co-operation of the USA (NASA), the CIS, Australia and other interested nations. This idea has interested several countries, because it is much cheaper to launch into space nearer the equator—but not so near that constant bad weather endangers each space project. A spaceport at Cape York Peninsular would be nearer the equator than its American equivalent at

Cape Kennedy, and less prone to hurricane damage. This proposal has been studied for the last ten years. Consequently, towns such as Cairns—which is presently a tourist, game fishing center—and Cooktown, should receive improved road and rail transport to facilitate the spaceport development. All this will take us closer to Gerard O'Neill's vision of space exploration (Chapter 7).

Telecommunications in Australia, under the government-owned TELECOM corporation, have been developed generally with the highest standards, and the opening-up of the "outback" by the use of solar-powered telephone/satellite links has made possible the use and population of otherwise inaccessible areas. People are not so cut-off as they were by great distances or low population density in the "outback" but can communicate with their neighbors. The greater use of facsimile (FAX) machines and automated telephone answering devices has given new life to small towns and isolated rural communities. In the next fifty years the average family in Australia will use computers (PCs) everyday for education, communication, business, and recreation. Improvements in telecommunications will allow wider use of modems to link one computer with another, making available vastly greater information to individuals.

Technology, particularly in the form of personal computers, will revolutionize education to the extent that each student will have his or her own PC and will carry fewer books, but more computer disks. The teaching profession will be shifted more toward aiding students to learn from prepared, interactive computer programs, which, when linked with interactive videos, will be able to teach each and every topic imaginable. A good teacher, therefore, will also need to be computer literate. Lesson preparation will become a whole new concept and teachers' training colleges will have to adapt to computers.

Students of different ability in different subjects will be allowed to progress at their own pace, guided mostly by the computer, but ultimately by their teacher. Some students will then excel in one or more subjects at a very early age, and the more gifted students will attend university before they are teenagers. Adult education, likewise, can be transformed by the use of computers, especially as voice-interactive programming becomes widespread. It is presently used mostly by the seriously-disabled, with liberating effect, but will soon find broader application in education and other day-to-day uses—even in the home.

Biomedical science is already experiencing a quantum leap in achievable results. Animal organs including donated human ones, can be transplanted to give new life to the chronically sick or injured. There seem to be no limits on what can be done with genetic engineering—except the moral ones—and so the next fifty years will transform our concept of health and medical science. Once the average lifetime of our citizens extends to 150 years or more, a major problem may be employment for a much longer period of life. Good health and fitness may extend the average age of retirement from 65 to 100. The accumulated experience of our older citizens can then be used more effectively—allowing greater time to pass on skills to the next generation. Young people will have to be highly educated if they wish to find a job after leaving school—since machines and robots will perform most of the laborious work. People will have plenty of leisure time, but this may be stressful if it is not used creatively.

The peaceful uses of nuclear power have been prevented in Australia by a myopic, and poorly-informed, sector of the environmental movement—probably because of the British postwar nuclear bomb tests which were conducted in the "outback" of Australia. Consequently, the average Australian has been "poisoned" against the considerable scientific, industrial, economic, and environmental benefits of the peaceful uses of nuclear energy—which could generate prodigious amounts of electricity for any purpose. Australia may need to utilize this technology for the restoration of the environment, and other reasons.

Nuclear power plants could be placed at river banks —where the water is either too saline, polluted with agri-chemicals, or too impure to drink—and used to pump it, purify it, and release it again, or store it for human consumption. Water could even be taken from the sea, desalinated, and pumped inland, for the benefit of isolated rural and city communities. The future development of many arid—and otherwise unusable—regions of the Australian interior depends on cheap, abundant energy supplies for irrigation, industrial and domestic electricity, and environmental restoration of the desert. The Northern Territory centered on Darwin, for example, could begin this new program of nuclear power installation.

All the problems of safety and reliability of nuclear power installations apparently have been overcome and can be solved if the nation has the will to do it. The Chernobyl accident was the result of irresponsible nuclear engineering practices, and the use of methods that would never have passed the safety standards of the West. It was nuclear power on the cheap, and unnecessary risk-taking. Such mistakes need never occur again, and the Australia of 2044 should be utilizing at least some of its vast reserves of "yellow cake" Uranium ore—25 percent of the Western world's total. The safe control of nuclear fusion (not using uranium, but only light elements like hydrogen and deuterium might also be achieved by 2044, in which case there will be even cleaner and environmentally, more compatible, reasons for using nuclear power to provide a better future for Australia.

CONCLUSION

One approach to defining the desirable values for the future of Australia might be to identify the commonly accepted, enduring principles of social and individual morality which have been the product of civilization over the last 5,000 years. Some of these principles, such as the Golden Rule ("Do unto others as you would have them do unto you") are obvious. Others might require more searching consideration.. On a shorter time span, the American and French revolutions of the late eighteenth century might suggest aspiring to "life, liberty and the pursuit of happiness" or "liberty, equality and fraternity." But in Australia, as in the western world generally, these same 18th century traditions are now in crisis.

Perhaps we could rely on restoring the best elements of the nonsecular humanist tradition, stretching back to Greece and Rome, such as respect for the individual coupled with duty to society (i.e., the formation of individual and civic character). The best elements of this tradition are also to be found, in varying proportions, in the great religious cultures of the world. The society we are aiming for is one not driven by consumerism, but one which is caring for both the individual and society, concerned about humankind and the world, geographically and environmentally. It is one in which a considerable measure of decentralization will allow the individual and the community to exercise influence. We are seeking a pluralist society, tolerant to differing ideas, but not plagued with a confusion of competing values. Therefore we must define and aspire to a higher culture.

In 1991 Barry Jones a former federal minister for science and a keen advocate of technological initiatives in Australia, identified nine major challenges:

(1) sustainable ecological development;
(2) energy efficiency;
(3) the future of cities;
(4) improving the quality of education;
(5) global problems from which Australia cannot distance itself (e.g. climatic changes, population growth, and poverty);
(6) health (including the problems of an aging population);

(7) making Australia "clever" (by reversing the decline in our cultural and intellectual infrastructure and strengthening ideology);
(8) restoring some role to parliament (by improving the quality of debate); and,
(9) coming to grips with complex issues (by improving the quality of public discussion). All these challenges need to be met in a spirit of service and desire to promote the common good—yet regrettably self-interest and survival are the negative factors likely to provoke action on them.

We have addressed many of these problems in this chapter. We have probably neglected the political crisis, marked by increased cynicism about political parties and government, by increasing lack of respect for the legal system, by reduced confidence in the police. If we do nothing about the future, if we allow present and likely trends to continue, Australia by 2044 will be marked by new features, some positive, some negative, some of indeterminate quality. These could include:

- the rise of third-generation immigrants to positions of power;
- continued economic deterioration;
- growth of decentralization;
- more responsibility for social services on the local community and the family (because of collapse of the welfare state);
- reduced democracy at the top echelons of government.

Tendencies toward decentralization and devolution are apparent already. This phenomenon is partly a political response to the feeling that the individual lacks power to influence the central authorities. If the centralized welfare state is unable to sustain itself, small local units may be needed to look after social welfare. Perhaps we will see a vast network of decentralized communities by 2044.

Yet within the decentralized structure, a centralized bureaucracy is likely to survive to attend to continent-wide concerns. The Green slogan, "Think globally, act locally," may yet achieve fruition. We must not end up with local city self-government, or worse: insensitive, authoritarian, federal government.

It may be that in order to introduce general reform some limitations of normal democratic procedures may be necessary, some "Special Law on the Economic Emergency" may be required. Otherwise, we must rely on traditional, time-honored methods: argument and debate; publicity; propaganda in books; journals; newspapers; and television. We must produce leaders prepared to assess new proposals and advance those which offer progress. In this respect, Australia has an unfortunate tradition of suspicion of outstanding leadership—that is, suspicion of so-called "tall poppies." Yet, this may be changing, as greater respect is afforded to community leaders such as former Prime Ministers Gough Whitlam, Malcolm Fraser, and Bob Hawke. Australia definitely needs leaders of ability and unquestionable integrity, who can guide the nation into the next millennium, and stimulate that level of renewal which can inspire the whole country to aim for noble goals, and sustainable development.

REFERENCES

1. P. Ellyard "Desirable Futures For Australia" in *Future* no.12, April 1989. John Eddy in *21st Century Magazine,* The Commission for the Future, 1990.
2. Geoffrey Blainey, *Eye on Australia,* 1991.
3. *Furious Agreement—Forty Prominent Australians Focus on the Next Fifty Years,* Australia, 1991. For Barry Jones, see pp. 248-251.
4. *Our Heritage and Australia's Future,* Australia, 1991.

5. Paul Hanman, ed., *Crosstalk,* The Australian Council of Churches: Boolarong Publications, 1991.
6. Alan Brown *et al.* "Trees for Community Needs: Australia's Major Contribution in Developing Countries" in J.N.Coles and J.M. Drew, *Australia and the Global Environmental Crisis,* 51.
7. Max Charlesworth "2000 AD: Terra Australia and the Holy Spirit," *Millennium,* Helen Daniel, ed., Australia: Penguin Books.
8. Alan Barcan "The School Curriculum and the National Economy" in *Issues in Australian Education,* J.V.D. Cruz & P.E. Langford, eds., Longman Cheshire, 1990.
9. *Education for What?* J.D. Frodsham, ed., Canberra: Academy Press, 1990.
10. John Hick, *An Interpretation of Religion,* Macmillan Press, 1989.
11. D. Millikan, *The Sunburnt Soul,* Lancer Books, 1981.
12. Department of the Prime Minister and Cabinet, "Ecologically Sustainable Development: A Commonwealth Discussion Paper," Canberra: Australian Government Publishing Service, 1990.
13. S. Dovers, "Sustainability: Definitions, Clarifications and Content," *Development,* 2/3, 1989, 33-36.
14. Ian Moore, "Water Resources and the Environment: Sustainability and the Need for a Transdisciplinary Approach" in *Second Ministerial Water Forum.*
15. H.A. Nix, "Water/Land/Life: The Eternal Triangle," Newsletter, no. 299 of the Water Research Foundation of Australia, 1991.
16. David Yencken "The Links Between Issue Identification and Action: Dilemmas for Scientists" in J.N.Coles and J.M. Drew, *Australia and the Global Environmental Crisis,* 253.

TWENTY- FOUR

THE CASE FOR JORDAN

Subhi Qasem

INTRODUCTION

Jordan is one of the smaller countries in the Middle East. In mid-1994, the population of the country reached 4 million and has been increasing during the last 25 years at a rate of 3.8 percent, a rate which is among the highest in the world. The high rate of population increase is due to natural growth of population as well as waves of Palestinian emigrants seeking refuge as they have been denied the return to their homeland Palestine. Situated in a central geographic position in the Arab region, Jordan has shared the Arab world in history, culture, aspiration and recent socio-economic and political environment.

Although the Arab countries are independent states, each one of them is highly influenced by events that take place in the region. If one wanted to address opportunities and dangers of scientific and technological advancements in the Arab region, he or she would have to do it in a regional perspective. In order to sketch a picture of different scenarios of how the Arab region including Jordan would be influenced by S&T, it is appropriate to summarize events and state of affairs as they have developed during the last 25 years and how they look now:

a) The Arab world includes countries that are among the richest in the world and those that are among the poorest. Jordan lies in the group considered to be among the lower middle income countries.

b) The Arab countries vary greatly in the way they are endowed with natural resources, human resources and capital. Sudan, for example, is one of the best endowed country with agricultural land resources but is among the poorest of the Arab countries if not the countries of the World. Egypt is well endowed with human resources but is also poor and suffers from economic difficulties. Jordan has very scarce natural resources but is well endowed with human resources that are among the best educated in the developing world.

c) During the last 25 years, the Arab world has witnessed wars and internal civil conflicts at a rate unprecedented in recent history. In 1967, 1973 and 1982 the region witnessed flareups of the Arab-Israeli conflict. The Iraqi-Iran war continued forcefully for eight years (1980-1988). Civil war erupted in mid-1970s in Lebanon and continued until the late 1980s. Civil wars and other conflicts erupted in the 1980s between Libya and Chad, Sudanese groups, Morocco and Polisario and Somalian political fractions. The Gulf war was among the worst developments that left the region with more uncertainties than ever before in recent history. Obviously, civil upheavals, wars between neighboring countries and regional conflicts have had their impact on the region. Jordan, like every country has suffered, although it was not directly involved in most of the conflicts. These events have not created a healthy environment for economic and technological development. On the contrary, the region suffered the most from one of the outputs of technological advancements: weaponry.

d) Three negative developments have taken place in the Arab countries during the last 25 years. The most critical is the continued widening of the gap between food production on one hand and the adequate consumptive food requirements of the population on the other. The degree of this food gap has varied among the countries. Jordan, for example, imported 60 percent of its food in 1991. Food imports consumed about 90 percent of total foreign currency earned from exports. During the period 1988-1990, food imports in the Arab countries have cost 5 percent of their total gross domestic production.

The second development is the decreasing per capita share of fresh water. This development is due to scarce resources in most countries and the slow development of water resources in others. The water per-capita share in Jordan in 1990 was about 190 cubic meters compared to 2.1 thousand cubic meters, the average for the Arab countries, and 9.3 thousand for the world.

The third development is the wave of emigration of Arab nationals to other countries within the region but more so to other regions of the world namely Europe, USA, Canada and Australia. The emigrating individuals vary in their level of education from unskilled to skilled labor in Europe and to professional in most S&T fields in USA, Canada and Australia. Some emigrants leave because of socioeconomic reasons. The majority of oil non-exporting countries suffer from a high rate of unemployment that reached about 25 percent in Jordan in the early 1990s.

e) The Arab countries including Jordan have been, on the whole, on the receiving end of technological output. They are not among the countries that determine global trends and priorities in technological advancements. It is true that Arab countries have benefited from global technological knowhow and output in their social and economic development schemes. However, many of the areas in the Arab countries that require technological breakthroughs have not scored high in the priority list of world leaders of technological systems. Deserts and desertification, desalination of water, solar radiation as a renewable source of energy are among the areas that require intensive effort both globally and regionally since they are of critical importance to the Arab countries. The Arab capacity to participate as an equal partner in the global system of S&T has been constrained by different obstacles over the last 25 years. How their role would be in the twenty-first century will depend on whether breakthroughs take place in solving problems that constrain positive participation. During the last 25 years, Arab countries have focused on building the infrastructure of S&T institutions. The expenditure on R&D as well as the ratio of R&D scientists among the labor force have been among the lowest in the world. The future development of these indicators may be important when future scenarios of S&T are considered. However, other socioeconomic and political elements are just as important.

f) The Arab countries have made substantial progress in building social and economic institutions to meet basic needs of their population. During the 1980s, 36 percent of total imports of Arab countries have been in machinery and instruments. These technological inputs were used in the establishment of facilities like those for health, civil aviation, telecommunication, housing, industrial and manufacturing in respective Arab countries. However, one of the major constraints of development that surfaced in all Arab countries was management. The most critical was the weaknesses in the information system and software technology required for sound management of renewable and nonrenewable natural resources as well as productive and service institutions.

OPPORTUNITIES AND DANGERS

How the Arab countries including Jordan will enter the twenty-first century will depend on development in each country and in the region as a whole. There are some developments that make one cautiously optimistic. On top of these are the efforts being made to bring about peaceful conclusions to one of the most important problems in the region: the Arab-Israeli conflict. There are encouraging signals from most parties involved in the conflict that promise to bring about a comprehensive and just solution. On the other hand, these signals may or may not produce the needed momentum in the peace process. Another encouraging development in the region and in Jordan in particular is the resumption of democratic institutions in some of the countries. Spearheaded by Jordan, democratic institutions are taking root again in several countries that include Egypt, Morocco, Tunisia and Lebanon. These developments, plus others, will determine whether the healthy environment will prevail in the countries of the region, a state that will determine the type and level of opportunities or dangers that technologies will bring about to countries of the region in the twenty-first century. Accordingly, three scenarios will be considered to explore the possibilities of technologies and what promises they may or may not bring to the region.

SCENARIO A

Continuity of status quo: Under this scenario the present conditions and development will persist with minor variations to produce basic changes. The opportunities and dangers under this scenario may be summarized as follows:

(1) Civil strife will erupt in more than one country in the region for the next 10 to 20 years, a situation that will interrupt true progress toward the sound evaluation of technology transferred from abroad as well as the implementation of efficient national programs to develop technologies that are congruent with sound development policies. The degree of social tensions will vary from one country to another. Some countries may enjoy more stable and secure conditions than others. However, the prevailing mood will be caution with very slow movement toward regional cooperation and integration of resources and efforts. Science and technology will not prosper in many of the countries of the

region. Such a state will be a reflection of the low priority and resources that will be allocated to S&T activities.

(2) The peace process between Israel and the Arab countries that started in 1991 will not gather momentum and will be interrupted because of weakness of confidence-building among the parties in conflict. The peace process will eventually break down within two to five years, an event that will start another race of building larger armies and armament in the region. The chances of another round of military conflict will increase with the most likely result that neither of the two parties will gain. During these years, more emphasis will be placed on technological R&D of military fields and their output. In some countries, not involved directly in the conflict, better conditions will develop that will allow healthier cultivation of technological opportunities.

(3) Cultivation of technological opportunities under this scenario will be limited. The region will benefit most from advances made in biotechnology in the field of food production. Several countries in the region, of which Jordan is in the forefront, are already using improved varieties of vegetables that are the product of biotechnological advances. As time passes, more biotechnology products will find their way to the productive system. However, since the Arab region will not be an important partner in technological developments, some of the pressing needs characteristic of the region will not receive adequate attention. Legumes, for example, are second only to cereals. These crops are not important in countries that will lead technological advances in the twenty-first century and hence the Arab countries will not benefit from technology output. Rice, corn and soybeans are not grown in large scale in the region. The areas presently planted with these crops will decrease with time due to shortages in irrigation. Water shortages are expected to prevail in the majority of countries of the region. Advances made in these crops will not be of great benefit to the region.

The technological advances toward desalination of brackish and sea water will be slow and will continue to be expensive. The countries will face water shortages close to crisis levels and will allocate most of the traditional water resources to municipal and industrial usage at the expense of food production. Consequently, the majority of Arab countries will divert more of their financial resources to food imports. These countries will have vast areas of land. However, land without irrigation under the arid and desert conditions of most countries will produce marginal output, if any.

Most countries in the region will attempt to cultivate advances made in the information systems to better manage their resources, productive systems and social institutions. The impact of the efforts given to this area will be limited because of drainage of resources and efforts to ensure internal stability and strengthen defense.

Some of the larger and richer Arab countries will attempt to allocate more resources to technological R&D. In a world that will grow in the twenty-first century to be the stage for competition among giants and larger groupings, these countries will not have great chances to achieve major breakthroughs because of the small size of S&T mass available to them. On the whole, Arab countries, including Jordan, will continue to import technological output. This may not be a negative development if these imports produce exports either in manufactured goods or services. However, the capacity of Arab countries to produce exports, other than raw materials or close to that, will be constrained because of their weak S&T base and insignificant R&D output.

SCENARIO B

Positive changes with some breakthroughs in some areas:

Under this scenario, conditions in the Arab countries will improve in more than one country or subregion. The following is a summary of the main feature of this scenario:

(1) Internal strife and social tensions will diminish in the 1990s opening the way for most Arab countries to enter the twenty-first century with more focus on solving social, economic and political problems. The environment that will prevail will be more conducive to the majority of Arab countries to allocate more resources to the proper evaluation of technology. The countries of the Arab region with Jordan among them will devote more resources to R&D and technology development. Regional and/or subregional cooperation and integration will be strengthened and countries of the region will work their way to become partners in the global system of science and technology.

(2) The Middle East peace process will continue in the right direction and will by the end of this century bring conclusion to the conflict between Israel and the Palestinians and Arab countries. The countries of the region will slowly work toward building bridges of confidence and consequently minimizing tension in the region. Most of the countries will devote less resources to armies and armaments that consume at present about 14 percent, on average, of GDP of countries in the region. More resources will be available for social and economic development and substantive improvement will take place in the overall living standards of the peoples of the countries in the region.

(3) More countries in the region will join in the establishment of sound democratic institutions, thus paving the way for the mobilization of national effort to positive activities including the sound choice and use of technology.

(4) The opportunities that technology promises will be cultivated in the majority of countries. Joint efforts in technology development will be linked to progress made in the degree of relation improvement among countries. Jordan, for example, being poor in natural resources and their quality will be able to participate substantially in the global system of technology development only if a greater degree of cooperation exists among countries of the region.

The country may be able to employ several technologies in the development of services that may be provided to the region. Biotechnology in medicine, food production and food processing will be used extensively by Jordan toward the twenties and thirties of the twenty-first century. Information systems to manage resources of productive and service sectors will become widely used. However, technology development and utilization on a large scale in fields like water desalination will be linked to how much countries cooperate regionally by pooling resources together and therefore be able to improve their chances for a breakthrough.

In the future, the countries of the Arab world will be able to play a more important role in technology development that answer their needs. The extent to which Arab countries will become partners in the global system of technology development will depend on how the present industrial and advanced countries will view the matter. If the big industrial countries move toward becoming less exclusive and allow more participation of other countries in dividing the "cake" of technology rewards, more and more of the small countries of the world, such as Jordan, will have a chance. However, if technology development continues to be more secretive and protected, most of the Arab countries including Jordan will continue to be customers of technology rather than partners in technology development.

SCENARIO C (IDEAL)

Under this scenario the Arab countries will overcome their internal and external difficulties and will move into regional cooperation and integration before the end of this century. Peace between Israel and the Arab countries will be concluded as the Madrid conference appealed for (within five years) and more and more Arab countries, including Jordan, will move toward integration with other countries, both economically and politically.

An Arab common market will be formed and several Pan-Arab financial and political institutions will emerge. Income generated from oil and other raw materials will be recycled within the Arab economies, producing a momentum of progress that will help speed the disappearance of economic disparities between the very poor and the very rich. The Arab countries as a group will form a strategy to enter as an equal partner in the global system of technology development. These developments in the Arab region will not be an isolated incident but will be part of a worldwide movement in which all countries and regions will participate.

The great promises of technology will be cultivated in areas that satisfy human requirements as well as the sustainability of its renewable resources. Nontraditional sources of water and energy will become not only a regional priority but a global one. Technology will make people come together rather than drift apart.

CONCLUDING REMARKS

Which of the scenarios will prevail is naturally the next question that comes to mind. Certainly, the ideal scenario is close to utopian thinking and will not come about in the foreseeable future. However, a combination of the first and second scenarios may be the more probable development which the Arab world will witness within the coming fifty years.

TWENTY-FIVE

POLISH BRAINSTORMING

Maria Golaszewska and *Tadeusz Golaszewski*

QUESTIONS

Predicting the future has been an activity of notoriety in our cultural history. The deciphering of dreams in ancient Egypt (for example, the dream about seven fat and seven thin cows and its interpretation as years of abundance and scarcity) and the futurology of the Greek prophets or contemporary scientists are historical facts. This problem of predicting the future has given rise to many questions.

Does predicting the future involve the creation of visions —whole pictures of the future world generated by an imagination that has been inspired by curiosity, anxiety and the hope of coping with yet unknown tasks?

Such visions have little rational justification; they are pure possibilities. So, maybe predicting the future has something to do with present reality. Maybe such predictions could be truths that are yet to be realized? But then, where does one's credibility come from, and moreover, where does that special knowledge come from?

Predicting the future might be defined as fortune-telling or interpreting omens about the future from contemporary artifacts. We can look at old folk customs, for example, fortune-telling from wax, where contracted marriages are foreseen.

There is, as well, a more scientific method of predicting —for example, meteorology that forecasts the state of atmosphere in a certain time by means of statistics, precise measurements, computer calculations and satellite observations. Futurology undertakes a similar prognosis in the sphere of social relations; for example, they foresee the numerical force of populations.

Visions of the future connected with religious cults have a quite different nature; the revelation of St. John the Divine, the most popular Christian prophecy, is a warning that foretells fatal consequences of immoral behavior. God and the inspiration of the Holy Spirit or revelations are the origins of those predictions.

Our task consists of creating unlimited visions of the future taken from the imagination and a discursive way of thinking, on the basis of brainstorming. Pre-vision then, is a kind of guessing, or intuitive searching for an answer to the question: what will the world be like in fifty years? We keep in mind the knowledge of developmental processes, social and historical trends. Assuming that every process in the human world has its own structure, taking these processes as a whole and thinking rationally, we can assess further stages of cultural and social development concerning a given population.

We must realize that the technique of brainstorming creates new awareness, not yet fully realized in scientific knowledge. Perhaps the most important value of such predicting is a moment of "self-verifying prediction." A man who has at least vague feelings of what will happen to him in the future tries to notice every omen in order to avoid the danger. This is connected to the hope that blind instincts and narrow aspirations could be dangerous to humanity as a whole, but they also may be used in order to avoid these dangers. Prognosis of ecological disasters arising out of the ecological movement is an example of such "self-verifying" warning predictions.

Our considerations are not of a universal character. They concern only the human world and are a description of a certain defined attitude, a report of the state of consciousness. This paper expresses an attitude of a contemporary Pole. At the same time we must remember that individual opinions are the result of cultural bias, education, ideology and even local experiences (such as the Chernobyl disaster that touched Poland) and resulted in ecological danger caused by pollution of rivers and of the atmosphere.

We cannot limit our considerations only to Poland's future: at the present, every nation coexists with others. We have our neighbors in Europe, and we are connected to other lands, for instance, America. The world situation touches us immediately through the great system of communication and information; inventions come to us and enrich our experience, and vice versa.

SENSIBILITY TO EXISTENCE AND IMAGINATION

Now we are in the middle of the year 2044. To describe the contemporary state of the world, we must mention the beginning of the twenty-first century and even the last events of the twentieth century. Very important discoveries and inventions had been made and they revolutionized the technical world and human life. The discoveries were stimulated by a total menace to humankind which, even if it existed before, hadn't been taken so seriously. That menace came from nuclear radiation and excessive industrial waste dispersed through the atmosphere. In the year 1997, just before the liquidation of the power station, the second disaster of Chernobyl happened. Windless weather made the nuclear cloud float high in the sky and a considerable increase of radiation went unnoticed. It grew slowly and constantly. And both major and minor damage around the globe occurred. We must add that improvements in power stations had scant effects.

Let's come back to the beginning of the twenty-first century: scientists tried to invent new sources of energy; two inventions were very important for the improvement of the world situation. For a long time scientists were interested in using solar energy but its fruits could be reaped only after they developed methods of storage. At that time, huge solar power stations began.

Poland participated in that enterprise to large extent. At the end of the 1980s, Polish society objected to the building of nuclear power stations in the country. But, because the building in Zarnowiec was far advanced, research on a new kind of energy processing began. An international group of scientists was set up and the first solar power station in the world started working. Because this discovery was so important for saving mankind from annihilation, Poland offered it to all countries which welcomed its use. A few years later, the UN forbade the building of other kinds of power stations and the older nuclear ones closed down. It was not a simple matter and sometimes it was full of conflict and complexity. Before they decided to send them into space, huge masses of radioactive substances had sunk to the bottom of the oceans. Although the idea appeared in the first decade of the twenty-first century, its realization took too long because new, massive containers which could absorb huge amounts of energy had to be constructed and devices which would send these containers into space as well.

Offering its inventions to other countries, Poland gained a lot. Because all our debts were released, many Polish experts were employed in rich countries and Poland started developing and getting relatively wealthy. Relatively—because nowhere in the world were people very well-off. Standards of living became almost equal because of the threat of total danger. People started to limit their needs and became thrifty.

Some of the most important inventions were made in the field of education. A demand for cultural unification, the possibility of fast communication and the transmission of information which penetrated all corners of the world were their inspiration. The methods of teaching foreign languages improved. The possibility of loading proper programs directly into the brain—into its particular sections—was a great success. (As the structure of the brain inspired construction of the computer, computer skills inspired this improvement of brain activities.) So people did not laboriously learn words or phrases of a foreign language, troubling themselves to understand, read or write. These skills were immediately obtained, thanks to suitable manipulations. Such methods were convenient for learning history, geography, medicine, and so forth.

So it could be supposed that humankind in the first years of the twenty-first century was delivered from many difficult problems, such as atmospheric pollution (which was supposed to be its greatest enemy). The air became cleaner, vegetation flourished, "black spots" vanished, the threat of thawing of the Arctic ice and of the submergence of parts of land disappeared. Living became easier.

RELIGION, ART, PHILOSOPHY

Common access to art, religious meetings and worship services diminished due to indifference. Faith faded away, churches became monuments of the past epoch. It happened even in Poland, where in the beginning of the twenty-first century religious life was relatively animated, as a consequence of appreciation and respect for the authority of the Polish Pope. Substitutes for religion and art reached Poland very slowly. But with the expansion of secular and political power of the Church, authentic religious emotions died. Churches, parishes and theological seminaries were deserted.

Spiritual leaders and people who kept faith and religious needs alive sought new forms of cult. People followed a particular religion beyond which there are some others that are "worse" or "false." The expanding ecumenical movement forced them to amend these convictions. Because of an integration of creeds, some of the exterior forms of the cult were no longer functional or adhered to by all the followers. On the one hand, people did not value all that is exterior. On the other hand, lack of these exterior forms reduced religious emotions. This caused a dilemma both for believers and theologians. In Poland (where exterior forms of the cult have been always valued) it was particularly difficult to convince the masses that the religious essence consisted of worshipping fervently with one's life, and behavior, demonstrating a clear confirmation of one's spiritual forces, moral conduct and contemplation: the grain of truth in all world religions.

For ages, patriotism and religions were traditionally connected in Poland, but in the fall of the former, detested communist regime (both anti-religious and anti-Polish) these two affairs were separated. New conflicts with our neighbors from the East who are as faithful and religious as we are had begun. Only clever tactical diplomacy and the threat of a complete disaster prevented a major military conflict (in the last years of the twentieth century).

But the patriotism of the Polish people changed as well. It became less invasive, tolerance deepened and pluralism of attitudes and values were at last appreciated. Cosmopolitanism was common throughout the world. Everyone was first of all, a citizen of the world and afterward, a citizen of the United States of Europe, which was founded in the beginning of the twenty-first century.

A new model of religious life was developed. True preachers of faith who did not aspire to gain political power and who gave up all material properties, appeared. A great center radiating spiritual forces developed that objected to the mechanization and degeneration of the world. Although living in different parts of our globe, priests and followers of different religions joined in a common effort for revival.

Using video-walls, they met at any time and they were able to consolidate their way of living. But the existence of these centers of spiritual forces and energies inspired many scientists in different fields to suggest a new way of world revival, a new order.

What happened to art? The artistic message of the twentieth century for the twenty-first century is conceptual art in which an idea is of essential value instead of a perfected work of art. An idea inspires the recipient to complete the work of art and give it an aesthetic value. This conception joined with computer technology leads to mechanical messages (instead of former works of art), audio-visual messages which would stimulate the individual imagination of the recipient and stimulate his need of artistic creation. At the same time, artistic expression approaches technical or computer games referring to creativity in its most general sense. The second message of the twentieth century is that art and life are intertwined: life is art and art is life. On the one hand, people appreciate the look of things, the beauty of town-planning and architecture, but on the other, a new kind of art conjoined with real life developed. There were elaborate detective games, played with some seriousness. People involved themselves in chasing and pursuing not only criminals, but also people chosen by chance. It was not done in order to punish them but for the pure pleasure of the game, for strong emotions and intellectual experiences. Solving a puzzle, for example, finding out someone's intentions and attitudes, foreseeing his behavior, etcetera, did not have anything to do with the desire of getting in touch with others; it became a pure game.

In 2044, traditional art remained common but on a rather limited scale, as the need for contemplation, for expanding the sphere of one's own experiences decreased. An all too frequent recurrence of the former art by means of mass-media caused dullness of sensitivity to classical beauty. Not direct emotions but structures and formal rules: hidden logic, consequence or lack of it, possibility of mathematical formulas, general principles was found in the former art. It was treated analytically (i.e., what postmodernism taught us in the end of the twentieth century) in the context of the whole culture and its changes. The reinterpretation of great masters of art happened and it was inspired by conceptualism.

In philosophy, two alternative ways of development appeared: on one hand, it became a counterbalance for increasing rationalization of all spheres of life, mathematization and computerization, both in science and everyday behavior. "A wave of irrationalism" observed at the end of the twentieth century turned people toward far Eastern philosophies and spirituality in which they found the way out from the blind alley produced by pragmatism, positivism, rationalism, or empiricism. On the other hand, the materialistic and mathematical world searched for scientific-philosophical reasons for its attitudes and habits. Philosophy came nearer to art as art has come closer to science and philosophy, and this approach was closer than it used to be at the beginning of conceptual art.

But it seems that both art and philosophy remained stagnant in new thinking and creativity. A new extraordinarily attractive philosophical system was needed to inspire the interest of many, and mutual respect for things without utilitarian purpose. This new perspective could deepen spiritual life, meditation and revival of feelings.

It may be that the warning of imminent human annihilation caused by atmospheric pollution enjoined the efforts of all nations to end wars and conflicts. So, too, the danger of the fall of spirituality and of the human nature, caused philosophy, religion and art to join their efforts to give the world energy for revival.

TO SAVE THE WORLD

In the beginning of 2044, the second Great World Conference on saving the world will be organized. Debaters will gather on four huge screens of the conference room —these screens will cover whole walls so that there will be an irresistible impression of personal presence of participants. They realize that the human world, if it is to exist any longer, needs something more than unpolluted air, a non-radioactive environment, etcetera. Since the first World Conference on saving the world was organized, a lot had been done. But people finally realized that the lifestyle of the contemporary world produced a huge desolation in human spirituality. It reviews the possible substitution of a wall-screen for direct interpersonal contacts, the substitution of pictures of things and nature for direct contact with them, and an exemption from personal effort in gaining knowledge. When art becomes a computer game, meaningless but interesting for sight, hearing and also for intellect (which stops integrating all spiritual human powers) it loses the capacity to inspire inner human forces. New kinds of art become anti-humanistic.

Many thousands of years will pass by the time the world reaches its balance. Is it possible to activate this process? Are we able to prevent the Earth from being crowded by freaks of unforeseeable features and further genetic transformations? It is surely possible if new energies, not only psychic but spiritual energies are set in motion.

Some people state that everything must start once again. There are some who did not fall into stagnation and they do not prefer pictures to reality. Many inventions from the twentieth and twenty-first centuries will be rejected because they make people resemble automatons. Interpersonal contacts which give a feeling of intimacy and coexistence must be renewed. People must be given a chance of laboriously looking for happiness in their own way.

Others reiterate once again the old warning of philosophers: complete rationalization and technicalization of the world led humanity up a blind alley, deprived of the sense of life. It appeared that greater sense of order is needed, and the more it is not appreciated and multiplied, the more of the sense of life is being lost. In the well-ordered mathematical world, where everything gets perfectly measured and weighed, comfortable, peaceful, where emotions and feelings have been eliminated, people fall into depression or suicide.

Does it mean that we must destroy civilization and return to life in the trees? No, this would be looking at externals, not looking at the essence. After civilization has been worked on for generations, the world is not the same and we cannot return to the beginning. Spiritual forces connected with biopsychic energies must be inclined not to the multiplication of technique and exemption of man from effects of his own creativity, but to the transformation of his mind, the release of feelings, the capacity of appreciating the nonrational.

WARNING VISION ONCE AGAIN

A most acute anxiety concerns the possibility of humankind's degeneration. The pollution of the natural environment combined with genetic engineering experiments may contribute to the appearance of a new species of human-monsters.

By the way, let us remember that the previous signs of such genetic mutation started in the second half of the twentieth century, before the Chernobyl disaster. The great Aral lake in Uzbekistan (64,000 km^2) was dried up. It was Stalin who decided that these waters would be useful to get the great wool plantation started. River waters feeding the lake were diverted for the plantation and the dried lake became so salty that all vegetation in it disappeared. Chemical products, applied to protect the cotton from disease, poisoned the water and the air. Great amounts of salt and chemical products rained down on territories where thousands of people lived. As a result, handicapped children were born, without hands, without feet, with lower atrophy. Salt coming from the dried lake was transported by winds to the Himalaya mountains, destroying soil and hastening the melting of glaciers.

So, is humankind to die without any cataclysm or evident "end the world"? Is the species of *homo sapiens* to become extinct as many other creatures have? Is it worth saving a species that condemns itself to annihilation? It is not easy to stop the momentum. Is it possible anyway?

RULES OF TRANSFORMATION

These are our conclusions:

(1) The year 2044 is not at all evident as a future fact of the human world. Humanity has sufficient means for self-destruction, destruction of the environment, and even the capacity to cause a cosmic catastrophe leading to the explosion of our globe.

Collective suicide is not taken into serious consideration because of the instinct of self-preservation, the will to life, and of individual and collective existence which—despite some aberration—still will dominate the human world.

But a deficit in social adaptation skills is already present. I am talking about acting against one's interest, against one's own individual or common well-being. If these behaviors are to increase over the next fifty years, complete destruction will occur in Poland. Such great effort of the will would be lost due to an impairment of the will, a lack of vision, and a feeling of general impossibility ingrained in Polish nature. This means that if a loss of motivation were to exist, it will lead to a feeling of total absurdity.

(2) The crucial element in the human world (especially in future perspective) is the axiologic factor of human behaviors, the regard for values which gives a deeper, metabiological sense to human life. If humankind is not to regress, a universal system of values must be created. So far, a system of primitive material values dominates. It is typical that the need of a comfortable, prosperous life and economic aims are more attractive than spiritual aims or even moral purity of

principles. The optimum model is a balance between vital and spiritual values.

In Poland, such values as faith or patriotism dominated while the economic and material sphere was considerably neglected for many years. Romanticism, after a turn to material values in the positivist stage, battled for sovereignty in the beginning of the twentieth century, and gained a new turn to the balance of values during twenty years between the wars. This was the communist ideology against which catholic groups were fighting. Poverty was evident in society and the desire for enrichment, the fast creation of prosperity really gave a challenge to important values.

It may be expected that in Poland the tendency toward reaching prosperity will endure.

The Catholic Church takes care of spiritual affairs in Poland. Now it is mainly interested in secular-political and Church-ceremonial affairs. Thus it loses its basic mission: the cultivation of the Christian life. If a new sudden effort of the collective will toward harmonious realization of the most important spiritual values does not happen, then stagnation in aspirations for principal values will prevail.

(3) A very important element of development is the recognition of the social-cultural situation, the knowledge about future aspirations of contemporary man. But a required component of this kind of knowledge is volition. The problem is to realize on a global and an individual scale what is correct, just, or necessary. The tendency toward short durations of endeavor is typical for Polish people: when violent fighting is over, relative stabilization follows, which is convenient for harmonious realization of development aims. But then a decrease in interest in public affairs sets in. The feeling "this is not what we fought for" appears, and indifference to the more important aims prevails. This is the reason that the strength of will decreases. Even if a theoretic, axiologic science is developed, even if ethical systems are established and a society creates intellectual elites, the lack of

efficient executive system (government, administration) will make social development regress.

The next fifty years may be a transition period, where the conditions for complete realization of Polish aims will be set. This may include exceptional achievements in science, creativity, art of labor management, but these achievements will be only single and occasional events which do not influence the whole of social life.

(4) Social disappointments are one of the most serious brakes on the harmonious development of the community. The situation of Poland is an example. In fact, as the communists stress, the police system eased, the Polish people gained greater access to money and goods, foreign travel became possible and many families enjoyed relative prosperity, many industrial investments were made, and so on. However, this success was only apparent. It rested on a large influx of foreign capital with high interest on loans. Finally, economic leaders of the state realized that Poland was threatened by financial breakdown and bankruptcy. The market ceased to be artificially stimulated, the controls on disposable income, drastic consequent caused disappointments, social upheavals, demonstrations, and, in the end, led to the birth of the "Solidarity" movement. But an even more profound disappointment happened when Polish people regained sovereignty, freedom and a democratic system. Poland realized how difficult it is to come to friendly relations with neighboring countries, to restrain excessive criticism of everything by everyone; to avoid excessive party division. Economic problems made the situation more complex. In this way, a stalemate was reached which undercut the most favorable changes: "This is not a Poland we fought for," recent enthusiasts of "Solidarity" claim.

Of course, it is possible to overcome these difficulties. They are only problems of development and transition. The faster the society will carry out contemporary tasks, the better will be a chance for a big step forward.

(5) Synergism of knowledge, energy and technique is one of the most powerful levers of progress. This synergism is produced by cooperation of many factors in order to achieve a common aim. A plane is a good example: its propulsion, streamlined shape and light construction contribute mutually to enable flight. This law of synergism concerns individuals and social groups as well. Individuals rarely can be gifted with high intelligence, talent for discursive recognition of reality, and possess at the same time a will strong enough to achieve this aim. This requires an extraordinarily high level of technological, practical and organizational skills in managing communities. It seldom occurs, but it has an incredible impact on society when it does.

Development of the optimum situation on a global scale depends on systems working synergistically.

(6) The main importance of all prognoses, predictions and future visions consists in inspiring, and consolidating hope that humankind has the possibility of further development. Hope creates the strongest motivation to act—even utopias may be useful to stimulate acting in certain directions.

If we take into consideration the many stresses, threatening dangers and conflicts that are still breaking out, a statement that people will survive till 2044 sounds a little utopian. If human beings are not sure that any future is to come, they will fall into apathy or into desire for debauched, primitive pleasures. But we ask: what will 2044 be like? And we want to find reasons that will convince us that in fifty years, many of our expectations will come true.

(7) When we are talking about survival and a development from a long-term perspective, we assume that an organism which is to develop is young, vital, and able to reproduce and improve. We can make an analogy with the life of a single person: the younger he is, the better the chance for development. So, is humanity young or rather old and decrepit? What can prove its youth or relative youth? Does science and technique develop dynamically? New centers of

spiritual transformation, new religions and moral systems begin (e.g., common care for handicapped people), may be a sign of youthful aspiration.

So, in the year 2044 humankind will yet be young. Some difficult tasks will have been carried out in only fifty years, and it is probable that human energy and the will of life will not be depleted so fast.

The definite end of humanity would come only when our planet gets as cold as ice. It may be that even a nuclear catastrophe would be overcome. But this is a remote perspective.

(8) But let's not deceive ourselves—retrogressive "development" is also possible. That means a decrease of the force of the will, diminishing interest in the most important values and the inward life. Vitality may weaken and resignation from the efforts for improvement may happen during the next fifty years. There may be possible retrogressive genetic changes in humans. What will be their consequences and how deep will they reach? Nobody knows.

This retrograde development need not be catastrophic. It may be only a readjustment.

9. Let us return to the question of Poland in 2044. We have suggested a possibility of breaking an impasse in which we found ourselves nowadays, but we also mentioned the danger of failure, disaster and even catastrophe. Let's consider certain moments that characterize the Polish turn to the future and hope for "optimal " development:

- After overthrowing communist totalitarianism, the period of transition to democracy, pluralism, liberty of conscience and capitalist ideas followed; this period, described as the "reconstruction period" carries manifold inconveniences, difficulties and dangers, which may stimulate collective effort;
- Slowly political elites and systems are shaped, including basic changes, the new constitution, and future perspectives, reaching beyond 2044;

- The commonly felt and urgently expected need for economic changes which can lead to stabilization, and harmonious cultural-social life are developed;
- Now in Poland everyone learns democracy, political culture, capitalism, new standards of intra-national coexistence, and international cooperation with Europe and the world community. (Let's take into consideration that the minister of Foreign Affairs was described as the most successful Polish politician;
- In the year 2044, Poland will reach the world level of a medium-sized stable and fully reliable country.

Reflections upon 2044 incite intellect and imagination mainly toward the need for radical changes—on the one hand to neutralize different threats, on the other hand to create new sources of enlivening spiritual and physical energy. Because hope still prevails over pessimistic despair, we conclude that humanity has remained young and can expect many wonderful adventures.

TWENTY-SIX

THE PHILIPPINES FIFTY YEARS HENCE

Andrew Gonzalez

INTRODUCTION

The Philippine hero Jose Rizal wrote in 1889 an essay entitled "The Philippines a Century Hence." Although the title hints at an early futuristic attempt, it is really more of a proposal of reform directed toward the colonial policy makers and administrators of the waning Spanish Empire in the late nineteenth century. Disappointing from the viewpoint of independence and nationalism is the assumption, on the part of Rizal, that the Philippines would continue to be a colony of Spain provided certain reforms were implemented.

In the intellectual community of Manila, the title has become a popular one when speaking of a futuristic view of the country; hence my choice of the title "The Philippines Fifty Years Hence," or the Philippines in 2044.

A hundred years after the publication of Rizal's essay, a group of academics in the Philippines organized a symposium to review what the Philippines had accomplished in a hundred years. Contrary to the dreams of Rizal, however, as enunciated elsewhere especially in his poem "My Last Farewell," the progress of the Philippines the past one hundred years has not been as glorious as wished for. In fact, it has been a disappointment.

THE PHILIPPINES MORE THAN A CENTURY LATER

The Philippines at present is a typical developing country, which has not attained the status of a Newly Industrializing Country (NIC) when compared to her neighbors, the tigers of ASEAN (Taiwan, South Korea, Singapore, Hong Kong) and a newly emerging NIC, Thailand, and most likely soon, Indonesia, though at present the per capita income of Indonesia still lags behind that of the Philippines. The vital signs of NIC status are more present however in Indonesia than in the Philippines, thanks to the former's consistent economic policy during the past twenty years, which is now beginning to bear fruit.

The Philippines has a per capita GNP of US$750, only about 52 percent of Thailand's US$1450, but still ahead of Indonesia's US$600, based on the latest annual report of the Asian Development Bank (for fiscal year 1990).

One of the major problems facing the Philippines is the ineffectiveness thus far of population planning programs because of cultural and religious reasons, so that the country has at present a 2.3 percent annual population growth, a life expectancy of 64 years, an infant mortality rate of 40 per 1,000 births, and a current population of nearly 64 million.

Even by optimistic middle-level projections of population growth, population is expected to be at least 72.5 million by the end of the decade (2000) and over 100 million by the year 2050.

There is an emerging middleclass of about 30 percent and an upper class of 10 percent, but 60 percent would still be considered below the poverty level. A more optimistic estimate by the outgoing Administration claims that the poverty level has gone down to 49 percent; the range will depend on one's definition of the poverty line.

The distribution of population between rural and urban is approximately 47 percent urban and 53 percent rural; the rapid urbanization (only 30 percent in 1970) has congested the urban areas, especially Metro Manila with a population of over 8.5 million and squatter families of approximately 38 percent; only about 40 percent of the population own their own homes, the balance renting from home owners or from government housing units which are leased for at least 50 years.

The annual growth rate in GNP at constant prices averaged about 6 percent from 1986 to 1989 but began a decline to the point where in 1991 there was a negative .5 percent growth rate.

Literacy (at least basic decoding) is 89 percent but functional literacy only about 75 percent; however, by and large, in spite of population increase and a poorly paid teaching corps, the elementary school level, a legacy of American colonialism, is in relatively good shape; however, the secondary school system needs radical improvement in staffing. Of every 100 children who begin in Grade 1 (the participation rate at this early stage is close to 100 percent), only 67 finish six grades, and only 36 finish secondary school (an additional 4 years). The tertiary school system, dominated by the private sector (85 percent), has over 1,000 colleges and universities, public and private, and graduates 12 of the 100 starters.

750,000 new employees in the work force are added each year from both secondary school and tertiary level graduates (250,000 per year), for whom jobs cannot be found.

This results in overseas work. About 2 million Filipinos are working in countries such as Hong Kong, Taiwan, Japan, South Korea, Singapore, Malaysia, Brunei, some countries of Africa, the Middle East, the United Kingdom and Western Europe either as technicians or as domestics, bringing in foreign earnings of over US$2 billion a year, more than each one of the main exports of the country (garments, electronic parts, coconut, and sugar).

The overseas workers are in addition to permanent immigrants, especially in North America; here about 2 million first generation Filipinos are now living and working.

The educational system at the tertiary level is a place where free enterprise has come to meet societal needs; however, because of market forces, making certain fields (e.g., commerce, engineering, and lately computer science) more in demand, there has been a neglect of fields even more urgently needed (pure science, vocational/technical, rural health/medicine). Moreover, because of the oversubscription of cheap nonlaboratory courses, and the great demand for expansion to meet a social need, Philippine higher education is plagued with a quality problem (the qualifications of the faculty, lack of library and scientific equipment) which makes the earning of a degree more important than its content.

The government system has been plagued with inefficiency and corruption, and a long bout with an authoritarian dictatorship (1966-1986) and an inept though personally honest administration (1986-1992) which has rendered the provision of basic social services a problem. The common problems of developing countries are therefore present: graft and influence-peddling in government, a loss of trust in law enforcement agencies, urban blight and congestion, poor transportation and communication facilities, poorly maintained roads, squatters and the lack of mass housing, pollution in urban areas and ecological problems because of much diminished rain forests (only about 800,000 hectares of virgin forests are left in 1991 from 1.2 million hectares in 1989 and 6.9 million hectares in 1976), erosion, lack of firm policies on land use resulting in the severe reduction of agricultural land for real estate urban development, a land reform program which has never really taken off and which perhaps is no longer viable under its present paradigm and needs thoroughgoing reconceptualization, insufficiency of power because of poor planning and implementation, delays in provision for needed power because of an inefficient central government, overcentralization of financial resources in the capital city, a large foreign debt which is making needed capital development almost impossible without either foreign aid or additional debts.

There is really no solid tradition of research in either industry and academia, most industries maintaining primarily quality control under "R&D" and really nothing else. Most colleges and universities are teaching institutions rather than generators of new knowledge, even about the Philippines.

However, medical education is at a relatively high level even by international standards although there is a tremendous egress of trained doctors elsewhere, staunched only by the reduction of demand for foreign doctors in North America. Adequate health services in the country are restricted to urban centers, especially Metro Manila, where world class hospitals and doctors are available. The imbalance however, results in poor public health services in the rural areas, to the point where there is still a significant minority of Filipinos who go through an entire lifetime without having consulted a properly trained doctor.

BASE LINE DATA FOR THE FUTURE

If one were to take a rapid survey of the projections made in the Study Papers for "Society in the Twenty-first Century: Opportunities and Dangers," certain projections are totally irrelevant to the Philippines in the sense that the infrastructural technology needed for the kinds of projections made is still nonexistent in the Philippines, given its present state of science and technology, as well as its lifestyle. On the other hand, because the Philippines is an over-schooled society, with world class institutions side-by-side in the same centers as typical Third World institutions, there are certain elements already present in the Philippines which can be taken for granted as given for further development and for projections of the kind cited in this book.

Given this mixed picture or patchwork quilt of achievements and failures, it is difficult to say if the Philippines really has a choice of futures since it is barely surviving. Before it will have the luxury of choosing alternatives, it has to provide for basic social services already taken for granted in the countries where the developments spoken of are possible.

The starting given in the Philippines is that science and technology are relatively underdeveloped in both industry and academia. Much technology is imported directly or at best adapted to local needs, but there is little or no research and development except in the area of food production through biotechnology, applied to rice strains and legumes and root crops.

The country is beginning to train many computer software and information technology workers. The beginnings of software production in cooperation with multinational companies which have a shortage of scientific workers are already in place for joint projects for First World countries done by Third World country technologists working in their home bases. What the washback effect of this kind of work means is not clear at present, although it would not be unrealistic to expect that a transfer of technology will be possible and the continuing exposure to the needs and problems of First World countries for software products will mean rapid technological transfer to the Philippines at least in urban areas.

Similarly, while robotics is nascent in the Philippines, it would probably not be fully functional in an overpopulated country that needs to employ as many of its people as possible to minimize emigration; the motivation to have labor-saving industries in favor of labor-intensive industries will not be there. This is not to deny the experience of developed countries where the presence of robots has not diminished the need for service industries. However, the initial impetus for labor-saving devices will not be in the Philippines for some time and hence might not lead to a rapid use of robotics unless of course the foreign market for services is so attractive that Filipinos will end up having no domestic help and therefore needing to provide themselves with labor saving devices for their households. This might explain the rapid increase in sales of washing machines in the urban areas at present.

The awareness of molecular biology is just beginning at present with its teaching in undergraduate courses and with the beginnings of molecular graduate research at the University of the Philippines. The advances, however, are more in plant biotechnology; scarcely anything has begun in research on human biotechnology.

In the field of energy research, where the Philippines has lagged behind because of poor planning and inept management, except for the discovery of significant oil fields in Palawan, and the transfer of technology (sea-based wells rather than land-based wells) that their extraction will call for, there is no other significant research in new areas, although there are the beginnings of teaching basic concepts in superconductivity in the physics departments of the main universities, thanks to the training of physics PHD's in Japan in these fields.

A conservation ethic is just beginning with the awareness of the dire results of ecological neglect and illegal logging, resulting in an almost total log ban as a policy and in a national replanting program led by the Department of the Environment and Natural Resources.

Any kind of ocean-based living in the future is out of the question although aquaculture and sea farming are now well established in the Philippines. Similarly, any talk of living in space would be totally alien to Philippine society at present.

For its energy needs, the Philippines will be depending on hydroelectric plants, fossil fuels, geothermal energy and thus far, little on solar power. Nonreliance on the latter is caused by the large initial expense involved, although solar energy has proven to be viable in the Philippines on a pilot basis using German technology.

In the field of rehabilitation medicine, except for physical therapy, there is no real research progress; small experiments of an interesting but noncommercial nature in the field of prosthetics and biomedical technology research have been done. For some time the Philippines will have to rely on imported technology for its advanced medical treatments, which are imported from North America largely through the contacts and frequent visits of Filipinos living abroad and of local doctors who return frequently to the United States to retool themselves since there is little of experimental work going on as far as medical research is concerned. Where significant medical research has been possible, it has been in the field of epidemiology and family and community medicine, rural health and preventive medicine, and treatment of the major communicable diseases rather than scientific experiments using sophisticated equipment, which is not available anywhere in the Philippines. However, the country, through its medical doctors, has been able to achieve world class status in various types of treatment and will continue this kind of technology transfer largely through the continuing contacts of Filipino doctors with North America. Hence, one foresees that the transfer of technology from North America to the Philippines will be quite rapid, at least for Metro Manila medical centers.

In the field of communications, the Philippines is quick to adapt and transfer technology both in transportation and in communications; one does not foresee significant discoveries, however, but merely borrowing. Of particular need in the country and therefore first to be imported will be any new vehicles which use fossil fuels more efficiently, cheaper types of vehicles (less polluting), and affordable means of rapid transportation (a train system which will be economically feasible to amortize). Unlike in India, however, there is no local industry of train building, but there is now a local industry for car parts manufacture and the extrusion of steel car bodies.

In the area of public transportation, reliance will be placed on buses, for which local bodies are now being made with the possible manufacture of large engines. A promising form of public transportation is the light railway system which has been proven to be functional in Manila. However, the costs of this are rather formidable; any form of less expensive above-ground transportation would be widely used not only in Manila but other urban centers such as Cebu and Davao.

REFLECTIONS

This short section makes some reflections on the development process as it is going on in a developing country such as the Philippines. What is the relation of this kind of slow development vis-à-vis rapid developments in fully developed countries which the different scenarios depicted by Morton Kaplan and Frederick Pohl, especially the one of Frederick Pohl, dramatize? Are these scenarios possible outside of North America, Japan, and Europe? We have examples in the other tiger economies of Asia and the Pacific, Taiwan, Hong Kong, South Korea, and most especially Singapore, which have managed in these two decades to break out of their centuries-old mold to join the modern, scientifically advanced countries of the world. There is thus hope, not necessarily for a utopia but certainly not for a complete dystopia, in countries such as Thailand, Indonesia and the Philippines.

One important question to ask is whether a developing country has to go through all the steps that developed countries have had to go through to reach their present level of development, including industrialization and its pollution and labor malpractices and a period of authoritarianism to jolt the citizens into a new awareness, and then rapid industrialization based on an export economy. To develop the way Singapore, Taiwan, Hong Kong and Japan have done, must countries such as Thailand, Indonesia and the Philippines go through the same steps or can they, under a revised geopolitical and regional structure in the area, by-pass some of these steps to attain a quality of life that is suitable for human beings in the year 2044?

Another point that must be made is that while countries such as the United States and those of the European community have a choice of scenarios, in many developing countries where survival is the order of the day, there is really no choice. In the Philippines, with its US$30 billion accumulated debt, there is no alternative at present but to aim for an export-driven economy to bring in the necessary foreign exchange to service its debts. Thus industrialization and the problems that go with industrialization by way of the pollution it generates and the heavy reliance it will have to place at least temporarily on fossil fuels or even nuclear energy make the price of development quite steep and open questions of "Is it worth it?" In actual terms, the question is academic since de facto through their economic managers these countries have chosen the route of industrialization and export-driven economic models to pay off their debts and to have the creature comforts (including energy-wasting comforts like air-conditioning) they have gotten used to.

Just as Japan industrialized, and to some extent imported Western technology yet maintained its core culture, other countries such as the Philippines may be able to maintain their inner soul without necessarily giving in totally to a Western worldview and value system, while imitating Western scientific technology and its ethos of work.

The Philippines, for example, in spite of rapid social change has maintained its folk religion and its adapted forms of religiosity, Christian and otherwise, in spite of the rapid importation of Western ways from the United States, its former colonizer. Moreover, the family unit in the Philippines continues to be a viable unit for care, protection, and security in old age, in spite of the fact that the Malay-type polygamous structure of the family de facto continues in an avowedly Christian monogamous society.

In the rapid social changes foreseen by the different scenarios, countries will participate in varying degrees as either contributors or users of the new technology. One foresees that the Philippines will be more of the latter rather than the former. And yet in spite of this, as the Philippines imbibes this new global culture built on science and technology and profits from its benefits and scientific advantages, it may, like Japan, maintain itself and its ways and absorb only those elements that are really compatible with its own world view and ethos.

As the country begins to grow in prosperity at least to the level of a semi-developed if not fully developed country by the year 2044 (perhaps by that time too the paradigm will have shifted making the distinction between First, Second and Third World countries and the dichotomy between developed and developing countries inapplicable and irrelevant), the Philippines will see a smaller population growth rate and fewer Filipinos migrating to other countries because of the availability of jobs locally and the improvement of the quality of life in this archipelago. In such an occurrence it would be interesting to see what the relations will be between locally based Filipinos and the Filipinos overseas, comparable to the Chinese *hua chaos* and the Jews in diaspora. What effect on the Philippines will the Philippine *hua chaos* have on their homelands?

NOTES ON CONTRIBUTORS

Marcelo Alonso, a physicist, is principal research scientist at Florida Institute of Technology, and editor of *International Journal on the Unity of the Sciences.* He was formerly director of Science and Technology for the Organization of American States. He is author of numerous books on physics and related topics.

Gordon L. Anderson is Secretary General of the Professors World Peace Academy (PWPA), associate editor of *International Journal on World Peace,* and lecturer in Religion and Society at the Unification Theological Seminary. He has published numerous articles on religion and peace, and edited, with Morton A. Kaplan, *Morality and Religion in Liberal Democratic Societies* (Paragon House, 1992).

Ben Bova is the author of nearly 8 realistic books about the future. His two latest novels are *The Trikon Deception* (co-authored with astronaut Bill Pogue) and *Mars.* He is President Emeritus of the National Space Society and current President of Science Fiction Writers of America.

Ernest G. Cravalho is the Edward Hood Taplin Professor of Mechanical Engineering at the Massachusetts Institute of Technology.

Christie Davies earned a Ph.D. in Social and Political Sciences at the University of Cambridge, in an attempt to understand Britain's American-style decline. He has taught at the Universities of Adelaide (Australia), Leeds (England, and Reading (England), and has been a visiting lecturer in India, Poland, and the United States. His work has been published in leading journals of sociology and public policy in Britain, France, Germany, India, and the United States.

Armando De la Torre, philosopher, is Dean of Social Sciences at Francisco Marroquin University in Guatemala. He is President of PWPA Guatemala. His paper was produced with the collaboration of: **Francis Aguirre**, technologist; **Ricardo Bressani**, biochemist; **Jorge Arias de Blois**, demographer; **Antionio Gillot,** mathematician; **José Asturias**, architect; **Bernardo Morales**, mathematician; **Gerladina Baca,** artist; **Luis Recinos,** psychologist; **Eduardo Suger,** physicist; and **Juan Carlos Villagrán,** physicist.

Ernest N. Emenyonu is Deputy Vice-Chancellor of the University of Calabar in Nigeria and President of PWPA Nigeria. His paper was produced in collaboration with **V. C. Uchendu**, Director, Institute of Public Policy and Administration, University of Calabar.

Maria Golasewska is Professor Emeritus and former head of the Department of Aesthetics in the Institute of Philosophy at the Jagiellonian University in Cracow, Poland. She is past and founding President of PWPA Poland. She has published over 180 works on philosophy, aesthetics, and literary and artistic criticism in Polish, English and French. She collaborated with her husband, **Thadeusz Golaszewski**, former head of the Department of Adult Education, Institute of Pedagogy at the Jagiellonian University.

Andrew Gonzalez, FSC, was president of De La Salle University from 1978-1991 and is now Vice-President for Academics. He is President of *The Manila Bulletin* and PWPA Philippines. He is a member of the Commission on the Filipino Language and the UNESCO National Commission of the Philippines, the Executive Secretary of the Linguistic Society of the Philippines and editor of the *Philippine Journal of Linguistics.* He is a professer at De La Salle University and holds a chair on the Philosophy of Education. A prolific writer, he has authored some 400 books, monographs, articles and reviews on such diverse subjects as theology, education, nationalism, phonology, and Philippine languages.

Morton A. Kaplan is Distinguished Service Professor of Political Science Emeritus at the University of Chicago, and editor and publisher of *The World & I.* He has published numerous books on International Relations, Political Theory, and Philosophy.

Jan Knappert, is Professor Emeritus of the School of African and Oriental Studies, University of London and President of PWPA Netherlands.

Hans Moravec is senior research scientist and Director of the Mobile Robotics Laboratory of the Robotics Institute at Carnegie Mellon University.

Gerald K. O'Neill (the late) was president of the Space Studies Institute in Princeton, New Jersey, which supports engineering research in space development. He has authored four books, one of which, *The High Frontier,* won the Phi Beta Kappa Science Book Award. He was professor of physics at Princeton University.

Frederik Pohl is a science fiction writer residing in Palatine, Illinois. His books include *The Space Merchants* (with C. M. Kornbluth), *Gateway, The Year of the City,* and *The World at the End of Time.* His current book, written with Isaac Asimov, is *Our Angry Earth* (published by Tor Books).

Jerry E. Pournelle is lecturer, consultant, social critic, and author of more than 20 books, including *Footfall, Lucifer's Hammer,* and *The Endless Frontier.* He is also an editor and columnist for *Byte* and *Infoworld* magazines. A past president of the Science Fiction Writers of America, Pournelle resides in Studio City, California.

Charles Sheffield is chief scientist and board member of Earth Satellite Corporation, former president of the American Astronautical Society, and former president of Science Fiction Writers of America. He is author of the best-selling nonfiction books *Earthwatch, Man on Earth,* and *Space Careers,* and more than 90 scientific papers, 80 short stories, and 14 novels.

S. Fred Singer has served in academic and governmental positions, most recently as chief scientist of the U. S. Department of Transportation. He currently directs a project on Science and Environmental Policy at the Washington Institute for Values in Public Policy. He is also Research Professor of Environmental Sciences at the University of Virginia, Charlottesville, and Distinguished Research Professor at the Institute of Space Studies and Technology, Gainesville, Florida.

Athelstan Spilhaus, oceanographer and meteorologist, invented the bathythermograph in 1938. He has been a physical oceanographer at the Woods Hole Oceanographic Institution since 1936, and was dean of the University of Minnesota's Institute of Technology for 17 years. He is now president of Pan Geo, Inc., in Middleburg, Virginia.

Claude A. Villee is the Andelot Professor of Biological Chemistry and Molecular Pharmacology at Harvard University. He has authored textbooks on biology, zoology, and human reproduction that have been translated into several languages, including Russian and Chinese.

Ivor F. Vivian is Associate Dean of the Faculty of Information Sciences and Engineering, Associate Professor of Mathematics and Chaplain at the University of Canberra, Australia. From 1973 to 1982 he was a member of the Australian Capital Territory Legislative Assembly and its predecessors. He is currently President of PWPA Australia. He collaborated on his paper with **Alan Barcan**, an honorary associate of the Department of Education, University of Newcastle, Australia, and **Patrick O'Flaherty** who has an extensive background in social welfare policy, and administration with the Commonwealth (Australian) Government.

Weerayudh Wichiarajote, is professor in psychology at Ramkhamaeng University in Bangkok, Thailand. He has been advisor to the prime minister's Department on National Development Ideology, and has been a specialist in Research and Evaluation Design for UNESCO. He serves as President of PWPA Thailand. He has published numerous articles and seven books on education, development, psychology and philosophy.

Alexander Zucker is Associate Director of the Oak Ridge National Laboratory, Oak Ridge, Tennessee. After a 20-year career in experimental nuclear physics, he has turned his attention to the development of high-temperature materials, including alloys, ceramics, and composites.